GOD PLANTED A GARDEN
(Horticulture in the Bible)

GOD PLANTED A GARDEN

(Horticulture in the Bible)

by
Dr. W. E. SHEWELL-COOPER,
M.B.E., N.D.H., F.L.S., DIP.HORT.(Wye), D.LITT.
Director of The International Horticultural Advisory Bureau
Chairman of The Good Gardeners' Association
President of The Farmers Christian Postal Service
The International Clans' Chief of The Campaigners
President of The International Miners Mission
Commandeur du Mérite Agricole (France)
Fellow and Doctor of The University of Vienna

Published by
ARTHUR JAMES LIMITED
THE DRIFT, EVESHAM, WORCS.
WR11 4NW

First Edition 1977

ISBN 0 85305 193 3

MADE AND PRINTED IN GREAT BRITAIN BY PURNELL AND SONS, LTD.,
PAULTON (BRISTOL) AND LONDON

To my Father
COLONEL E. SHEWELL-COOPER, R.A.
First President of the Officers Christian Union
and the
First President of the Crusaders Union
and
who introduced me to the Bible

ACKNOWLEDGMENTS

The author's thanks are due to the late Rev. Dr. Fred T. Ellis for urging that this book should be written, and for reading through the text and making certain suggestions. Also to Dr. F. Cawley, the late Prebendary Colin C. Kerr and Canon T. L. Livermore, for kindly encouragement and finally to Miss Betty Tyler, the brilliant Christian artist for her illustrations.

CONTENTS

CONTENTS

CHAPTER 1

THE BIBLE BACKGROUND

IT MUST BE remembered that the Bible was originally written in two languages, Hebrew in the Old Testament and Greek in the New Testament. It deals with two main periods, the sojourn of the children of Israel in Palestine until they were taken into captivity, and then there is a space of four hundred "silent years", as they are called, until the New Testament starts its teaching with the coming of the Lord Jesus Christ. From then on we get the picture of the pathways trod by the Lord Jesus, the journeys of St. Paul and the apostles, the letters of instruction and encouragement that were written and finally that wonderful prophetic book—the Book of Revelation.

When, therefore, we are studying the plants of the Bible and writing about them, we are dealing with the period from the beginning, when God made the world and peopled it—as well as with a comparatively short period from the time of Christ onwards. The bridge of time which spans the interval between the Old and New Testaments is probably from 400 B.C. to 4 or 5 B.C., for the Bible does record events immediately preceding the birth of Jesus, like the announcement to Zacharias, and the Annunciation to Mary.

Those who were responsible for the translation of the Bible into Latin, in the case of the Vulgate (which was completed in A.D. 400), have of course tried to name the plants correctly, but there have been in some cases many differences of opinion.

Wycliffe did his work largely with the Latin Vulgate. Tyndale, who was a Greek scholar, turned largely to the Greek translation of Erasmus. His New Testament was first issued in 1525. Most of us today use the Authorized Version, which is often referred to as the King James Version as it was prepared on the instruction of King James I of England. It was made available in 1611 and it is this particular Version of the Bible that the author has largely used in his studies for this book. He has also made use of the Revised Version which was made available in 1884, and is extremely useful because it touches the ancient copies of the original Scriptures which were not available in 1611.

The two books of the Bible that have most references to agriculture and horticulture, interestingly enough, are Genesis and Isaiah. Almost every book of the Bible, however, refers to plants, except the Book of Lamentations, the Epistles to the Philippians and Colossians, the Letters to Timothy, Titus and Philemon, and the Book of Jude. There are very many references in the first five books of the Bible: fewer references in the twelve historical books from Joshua to Esther, far more again in the five poetical books from Job to the Song of Solomon, a very large group in the major prophetical books, that is, the five from Isaiah to Daniel; and little reference, on the whole, in the books of the twelve Minor Prophets from Hosea to Malachi.

Matthew and Luke have the greatest number of references in the New Testament though there are many in all four Gospels. There are some references in the fourteen Pauline Epistles, but hardly any in the seven General Epistles. There are about twenty-five references to either agriculture or horticulture in the Book of Revelation.

CHAPTER 2

THE GARDENS AND GARDENERS OF THE BIBLE

TO HAVE GARDENS it is usual to need gardeners and in Great Britain it is estimated that there are twenty million keen amateurs. The Greek word *kepouros* is mentioned only once in John xx. 15 when Mary, seeing the risen Lord Jesus, supposed him "to be the gardener". Of course, speaking spiritually, this is His function, to tend, feed and water His "plants" and to expect them to bear one hundredfold.

Adam was a gardener and his son Cain was a good vegetable grower. As a trained gardener myself, I believe that the questioner in Luke xiii. 8 must have been an expert horticulturist because he knew that the tree which had become unproductive should be organically manured.

Solomon, I feel, was a keen amateur gardener for his writings betray his thoughts. Paul, though officially a tent-maker, was a knowledgeable gardener when he talks about grafting and the right stocks for fruit trees. Isaiah may have been a keen amateur gardener because he writes about taking cuttings in Isaiah xvii. 10.

We read in Genesis, chapter ii. 8, "The Lord God planted a garden eastward in Eden; and there he put the man whom he had formed." This garden grew the most beautiful trees, for we read that they were "pleasant to the sight" as well as being "good for food". It was certainly a watered garden, for it was literally surrounded by water, rather like an island, for we read "a river went out . . . to water the garden" (Genesis ii. 10).

FIG. (Ficus Carica).

Adam must have been delighted with the ample irrigation provided, when, as verse 15 says, "The Lord God took the man, and put him into the Garden of Eden to dress it and to keep it."

Those of us who have worked on the ground in Spain and in parts of Africa know the tremendous value of water in hot regions of the earth. A watered garden in Spain is indeed productive, and the Spaniards during the last few years have been doing everything they can to sink wells and thus to provide artificially the necessary vast quantities of water for irrigation. The Garden of Eden had water provided for it, and the Easterner cannot think of a good garden without water. In fact in my book "The Basic Book of Rock Gardens and Pools" I mention the Persian motto which says, when referring to a garden pool, "If there is a heaven upon earth—it is here—it is here!" In all Indian and Persian gardens water is necessary, and if we remember that the Bible is primarily an Eastern book we can perhaps envisage the great part that water plays in the cultivation and beauty of a garden. It is really to the East that we owe much of our love of the garden pools found in British gardens today.

Surely, the main requisites of the Palestinian garden were first of all water, second shade, and third scent. Water, which was indeed life to the plants and joy to men, shade in the heat of the day, and scent which was so beloved by the mystic Easterns. When considering this Garden of Eden we find that nothing is said about flowers, but the stress is upon trees. These would produce not only plenty of shade, but also scented blossom followed by delicious fruits. Far more is said in the Bible about trees than flowering plants.

Great interest has been aroused in the few garden

scenes that we are allowed to envisage in the early chapters of Genesis. There is the walking of God in the garden, in the days before Adam and Eve had sinned, so that they were able to walk and talk with Him after they had done their work each day. It is indeed a lovely term "in the cool of the day", which we read in Genesis iii. 8. The Authorized Version has it, "they heard the voice of the Lord walking in the garden", but to me the Hebrew word used there suggests "sound" rather than "voice", and the translation should therefore be, "they heard the sound of the Lord God walking in the garden". How tremendously awe-inspiring that is! It is no wonder that having so seriously sinned by stealing, lying, and deceit, they immediately hid themselves in the dark shade of the beautiful trees.

The last picture that we get of the garden is equally awe-inspiring: "And he placed at the east of the Garden of Eden Cherubims, and a flaming sword which turned every way, to keep the way of the tree of life." There was no re-entering into that luxuriant eastern paradise, where there were shade, shelter, peace and water, where there was perfect communion with God, and where life had no end. There is an interesting reference to the garden, however, in Ezekiel xxxi. 8, 9: "The cedars in the garden of God could not hide him: . . . nor any tree in the garden of God was like unto him in his beauty. I have made him fair by the multitude of his branches; so that all the trees of Eden, that were in the garden of God, envied him." This is referring to the King of Assyria, but it does show what great importance was placed on the trees of Eden—their many branches, their elegance, and their beauty.

We should remember that many of the ancient Eastern folk, especially the Egyptians, Israel's near neighbours, made gardens not only for their enjoyment while on earth, but also that their spirits might rest in them when they were dead. There was a tradition that men's spirits came out of their tombs "in the cool of the day", and enjoyed the shade, and the water of a beautiful garden.

No one, of course, knows where the Garden of Eden was, but in Medieval times it was thought that it must have been an island off the coast of "the Eastern Ocean". An age-old belief was that it was part of the sunken continent of Atlantis. In the seventh century, Cosmas said it was a region to the east of China, watered by the four great rivers mentioned in Scripture, each of which sprang from subterranean sources around the garden. There is a Saxon version of the Latin poem, "De Phoenice", which describes paradise as an island in the East where there was no suffering and no night, and where the trees bore their fruit throughout the year, and where the air was filled with the sweetest of odours. Even the Red Indians have their own thoughts about the paradise where the streams are full of good fish, and where the hills are covered with fruit trees laden with delicious fruit.

There are no laws concerning gardens in the Bible, unless one includes the one which refers to the removal of your neighbour's landmark (Deuteronomy xix. 14). Actually this undoubtedly refers to farms, but one could equally well apply it to gardens. On the other hand, there are laws concerning trees, and this is one of the reasons why one can say that they were so important. Take, for instance, Leviticus xix. 23. It is quite clear that no fruit was to be picked for the first three years, that is to say the children of Israel were to encourage the trees to grow by pruning, and they were not to ruin them by trying to make them crop too early. In the fourth year all the fruit picked was to be offered to God, and in the fifth year the gardener could pick the fruit for himself.

One can turn over further to Leviticus xxvii. 30 in order to discover that a tenth of the fruit borne by trees was to be considered holy to the Lord God. While in Deuteronomy xx. 19 we see how God places the greatest importance on trees, for he says: "Thou shalt not destroy the trees thereof by forcing an axe against them: for thou mayest eat of them, and thou shalt not cut them down (for the tree of the field is man's life)". Verse 20 reiterates this point very clearly. We can therefore say that the gardens of the Bible are largely places where trees grew happily to provide shade and food and where water abounded to give coolness.

Furthermore one should make it clear that all the gardens mentioned in the Bible are definitely associated with wealthy people. Just as you do not get gardens today in Spain in front of, or at the rear of, the houses of the poor, so you do not in the East. The garden in which our Lord was buried belonged to rich disciples. Let us, therefore, think of the Biblical gardens as a luxury of the wealthy, and not as in this country the hobby of millions of people. The typical luxurious Spanish garden, for instance, is one where water runs straight down the centre, or where there is a long series of narrow, deep, fascinating pools, probably due to the Moorish influence.

There were market gardens in Egypt, and Moses makes reference to them in Deuteronomy xi. 10: "For the land, whither thou goest in to possess it, is not as the land of Egypt . . . where thou sowedst thy seed, and wateredst it with thy foot, as a garden of herbs". Few people know what is meant by the term "with thy foot".

The main channel of water would run down the centre of the garden, and there would be shallow furrows running between all the rows of crops. It was then a simple matter for the gardener with his bare foot to move some of the soil from the bank of the main channel so as to allow the water into one of the furrows. When this particular furrow had had sufficient of the life-giving water, say in half an hour, he could dam up with his foot the opening he had made, and then break down the soil higher up so as to let the water into another furrow. I have watched gardeners in Madeira, Corsica, and in Spain, do this again and again.

When referring to the garden of herbs in Deuteronomy xi, we may consider it to be a kind of chessboard garden where all kinds of vegetables were grown: leeks, melons, cucumbers. There is little doubt that the Israelites copied the Egyptian idea, and so in very early times Palestine was irrigated with water by this method. Once again Genesis xiii. 10 emphasizes the importance of water to the Israelite gardener for when Lot had to make his decision he definitely chose the plain of Jordan because it was "well watered everywhere . . . even as the garden of the Lord, and like the land of Egypt".

One might almost have called this chapter "the water of Palestine", rather than the gardens, because we come back to the importance of water again and again. When Balaam was making his prophecy in Numbers xxiv. 5, 6 he said these words: "How goodly are thy tents, O Jacob . . . they spread forth as gardens by the river's side, as the trees of lign aloes which the Lord hath planted, and as cedar trees beside the waters. He shall pour the water out of his buckets, and his seed shall be in many waters." This system of lifting water by buckets attached to wheels is still common today in many undeveloped parts of the world. The donkey walks round and round in a circle, and this causes the big water wheel to revolve and the regular series of buckets tied to it to pick up the water from the well and throw it into the channel down which it runs to the garden below.

It is very difficult to know whether "groves" ought to be included under the heading of gardens. These are often mentioned in the Bible as being places where sacrifices were made. Usually these special gardens were condemned by God as in Isaiah i. 29: "Ye shall be confounded for the gardens that ye have chosen." It is obvious here that the reference is to the sacrificial groves which the children of Israel made again and again in which to make offerings to heathen gods.

In the early days it was obvious there was nothing wrong in making an attractive garden or grove in the name of the Lord, and there are definite indications that a suitable grove, hedged about, could be, as it were, an open-air church or assembly. Abraham, in Genesis xxi. 33, "planted a grove in Beersheba and called there on the name of the Lord, the Everlasting God". As he went on living in that area for a fair length of time, it seems obvious that he used this garden as a regular meeting place for prayer and praise.

Later on it would seem from the laws of Moses that the children of Israel had been badly influenced by the heathen groves where evil practices took place. It was a question of providing a secret place in which men and women might sin without being observed. This led undoubtedly to the instruction, "Thou shalt not plant thee a grove of any trees near unto the altar of the Lord thy God, which thou shalt make thee" (Deuteronomy xvi. 21).

Ahijah, the great prophet of the Lord, was very definite in his condemnation of the gardens or groves which Israel had made (1 Kings xiv. 15) when he spoke to the wife of Jeroboam and said, "For the Lord shall smite Israel . . . he shall root up Israel out of this good land . . . and shall scatter them beyond the river, because they have made their groves, provoking the Lord to anger." Thus the making of groves was regarded as a tremendous sin. Ahab, of course, made a grove, and this was denounced in 1 King xvi. 33; Gideon's father made a grove to Baal and this was spoken against in Judges vi. 25 and in fact Gideon was told definitely to "cut down the grove".

There are people who say that because many of these groves were planted in high places the influence spread from the Chaldeans right through the ages to Elizabethan times when mounds, groves and temples were a feature of the big gardens of that day. If groves had been used, as it is claimed, in the time of Ur of the Chaldees, then Abraham would have known them, for they had been planted and used some three hundred years before his time. It may have been this influence that caused him to plant a grove unto the Lord at Beersheba.

Whether or not these groves were always in high places, Isaiah makes no bones about condemning them, for speaking in the power of the Spirit he says in chapter lxv. 2, 3, "I have spread out my hands all the day unto a rebellious people . . . a people that provoketh me to anger . . . that sacrificeth in gardens, and burneth incense." There is little doubt, therefore, that God hated these special gardens made for evil practices, and for sacrifices to other gods.

There were, of course, luxurious royal gardens such

as are described in the story of Esther. Whether these were courtyards rather like the quadrangles of the Cambridge Colleges, we do not know, but the Bible says that there were white, green and blue hangings fastened with purple linen cords with silver rings round the marble pillars. The pavements were of marble also, and the beds are described as being of "gold and silver". It may have been that these courtyard gardens led out into the park that adjoined where the king would do his hunting, for the parks of those days were laid out specially for that purpose.

One often wonders whether the Jews who were taken captive by Nebuchadnezzar had to take part in the construction of the hanging gardens of Babylon which, interestingly enough, are not actually mentioned in the Bible. Anyway, Nebuchadnezzar certainly needed the craftsmen, for he carried away one thousand carpenters and smiths who would be very useful in helping to make the hanging gardens, or at least to keep them in order. These special gardens may have been very beautiful, and certainly were unique, but one wonders whether they had the richness and perfection of the gardens made by that wisest of all kings—Solomon.

He seems to have been a great plant-lover, and he made collections from all parts of the then-known world. His marriage with the great Egyptian princess probably influenced the making of the gardens, for he would have seen in her country the wonders of the horticultural planning of her father. Gardening was a labour of love to Solomon, as it is to all of us who have "green fingers", and Ecclesiastes makes it clear in chapter ii. 5, 6, 10 that his heart rejoiced in all his labour of garden making. Let us quote his own words: "I made me gardens and orchards, and I planted trees in them of all kinds of fruits: I made me pools of water, to water therewith the wood that bringeth forth trees . . . And whatsoever mine eyes desired I kept not from them . . . for my heart rejoiced in all my labour: and this was my portion."

Solomon evidently had a nursery where he raised his shrubs and trees, for in chapter ii. 6 the Revised Version intimates that the translation should be "pools of water, to water therefrom the forest where trees were reared". He evidently made garden upon garden, and watered them from vast pools. The great Jewish historian Josephus tells us that the Royal Gardens abounded with rivulets, and even today the squarish-looking hill at the side of the valley of Hinnom is called by the Arabs "The Mountain of the Little Paradise", a term which undoubtedly refers to the ample water provided for those beautiful Royal Gardens.

The present-day herb gardens, of course, are devoted largely to culinary herbs like sage, mint, marjoram and thyme, but the Eastern herb gardens were planted with sweet-scented shrubs as the Apocrypha points out, and Solomon evidently concentrated on myrrh, spikenard, cinnamon, frankincense, aloes, and saffron. The camphire mentioned is the henna used by the Arabs today, while the spikenard is related to the Red and White Valerian found in our gardens. It was this spikenard that was gladly given as a very precious ointment to our Lord in Mark xiv. 3. The Egyptians knew the value of using alabaster vases and boxes for perfumes because the odour could exhale through them. Scripture tells us that Solomon sought far and wide for his scented perfumes.

There are those who believe the scent of many plants has been lost, and they always give as an example the musk. Whether this is due to general degeneration, or to subtle virus diseases, we cannot say. It is obvious, however, that fragrance played a tremendous part in Solomon's garden, as well as in the gardens of other Eastern potentates. Even the newly opened vine blossoms were said in those days to have an exquisite scent,

MYRRH (LADANUM)

which they certainly do not have today. We use the word "flower" to cover in a generic sense the blossoms of various plants, but the Hebrew language only knows the word "*bosem*" which really should be translated "scented plant". It would seem therefore that to the Old Testament folk all flowers were fragrant, whereas today they certainly are not.

In the Song of Solomon vi. 11 we read "I went down into the garden of nuts". This should have been an

ALMOND (AMYGDALUS COMMUNIS)

interesting garden, and it obviously grew almonds, for these were known in Palestine as early as Genesis xliii. 11 and Exodus xxv. 34. Some think that Pharaoh was offered pistachio nuts by Jacob because these did not grow in Egypt and therefore were looked upon as a luxury. Walnuts have certainly been known for thousands of years, and Solomon would probably have included them, and possibly hazel nuts and chestnuts, in his garden, for both are mentioned in Genesis xxx. 37. Some people argue that the Hebrew word for "chestnut" should be rendered "plane"; one can riposte, however, by saying that in Pliny's time many varieties of

CHESTNUT (PLANE TREE). *Platanus orientalis.*

chestnuts were known, so there seems no reason why King Solomon should not have included some of them.

Because shade was so important, arbours were made in gardens, and it is probably to these that 1 Kings iv. 25 refers: "every man under his vine and under his fig tree". Micah iv. 4 repeats the promise as does Zechariah iii. 10. It was up the trelliswork and over it that the vines and figs would be trained so as to give shade. Such a booth was temporarily constructed by Noah, and it was probably under such a shady trellis arbour that

THE VINE (Vitis vinifera).

Nathaniel was sitting when our Lord's spiritual eyes saw him. People in the hotter parts of Europe often train their vines along wires or trellises over the courtyard today so that they can sit in the shade and have meals there. It is said that the gardens of Jerusalem had to be outside the city walls because the usual way of feeding the soil in those days was by human manure.

As a gardener I like to think that our Lord loved gardens, for He certainly frequented the garden of Gethsemane. We read, for instance, "Jesus ofttimes resorted thither" (John xviii. 2). He was buried in a garden and as Joseph of Arimathea was wealthy it was probably a very beautiful one. Furthermore, Joseph evidently kept one or more gardeners, for Mary in her ignorance, and with tear-filled eyes, failed to recognize her Master and thought He must be the gardener.

CHAPTER 3

THE FLOWERS IN THE BIBLE

THERE MAY BE some reading this chapter who will remember Mrs Ewing's children's book "Letters to a Little Gardener". Here she says, "Don't let me forget to pray for travellers when I thank God that I am content to stay in my own garden. It is indeed furnished from the ends of the earth." We have to thank the intrepid botanists who have been all over the world seeking for beautiful flowers and have brought them to this country, where they may give delight to those who grow them.

Today, we are most careful to give our flowers Latin names, so that whether we are in Japan, or South America, we know exactly the plant about which we are writing or talking. Unfortunately, however, in the olden days, flowers were given local names, and sometimes these meant one plant in one district, and quite another plant in another county. As it is possible for such things to happen within an island like Great Britain how much more so can it take place from country to country. When trying to make a study of the flowers of the Bible, one is up against this problem.

Having been to Palestine I can tell you that one of the great joys of the country is the wealth of wild flowers which are found not only on the plains, but also on the mountains. It is said that there are over five hundred species in Palestine which are grown in Great Britain, and probably yet another five hundred species which are indigenous to Palestine alone. Our Lord, as He preached to the multitudes, must have stood on mountains and on plains which were literally carpeted with wild flowers of all kinds, and it is no wonder that the Psalmist cried, "The works of the Lord are great; sought out of all them that have pleasure therein. His work is worthy to be praised, and had in honour." This is the grand prayer-book version of Psalm cxi. 2, 3.

We who love flowers can certainly see God's works in the countryside, for He has made the minutest flowers so perfectly that it is really worth while learning about them and "having them in honour".

It may be as well at this point to take the flowers of the Bible one by one, and to see what we can learn about them and from them.

Desire

The word "desire", referred to in the passage that is so often read at funerals—"the almond tree shall flourish and the grasshopper shall be a burden, and desire shall fail"—apparently refers to the flowering plant *Capparis sicula.* It is a type of caper and may be compared to the capers that are bought in bottles today for use in sauces and as a relish. These caper berries are eaten as a kind of hors d'oeuvre in Palestine with the idea of giving the diner a desire for food. The caper berries are found around Jerusalem, on the hills round about Nazareth, in the valleys of Sinai and, I believe, even in the Wilderness of Judaea.

The actual Hebrew word used for "desire" is *tapher,* and it has been translated in the various versions I have read as "caper", "caper berry", and "caper tree". The plant itself trails, bearing long horizontal leaves at regular intervals. It flowers during the month of May and the blooms are often three inches across, being pure white. These flowers are filled with pretty rose-coloured stamens tipped with yellow. The attractive buds are picked off before they open and pickled in vinegar. It is claimed that these will give even old men a desire for food and it is to the young flower buds, therefore, that the text refers. If this is so, then Ecclesiastes xii. 5-7 probably indicates that the man is so old that even the stimulus of the caper will not make him want to eat or drink, and that is why he is returning to the earth as humus. It has been suggested that because the berries are borne on short spiny stems and these tend to droop when laden with the weight, there is a picture here of an extremely old man bowed down with his grey hairs near the grave! Another meaning of the word "desire" is that it is an aphrodisiac. If therefore the berries failed to have a stimulating effect on men—then they really were old!

I am told that even today there are certain islands in the Mediterranean where the buds of this plant are still gathered and bottled in vinegar. On these Greek islands the word used for "caper" I understand is still the same word for "desire", and if this is so, then this is an extra pointer to the meaning of this text. The plant has a trailing habit of growth and is rock-loving.

Lily

It would seem that the lily of Song of Solomon ii. 2 was the Madonna lily which was planted in those days for medicinal reasons. Of course one cannot be sure, but there is little doubt that these Madonna lilies were quite plentiful in Biblical times, and it was only when soil erosion took place owing to the cutting down of the forests that these lilies gradually disappeared. They are growing again quite successfully in Palestine, however, and I very much admired them when I was last there. The Madonna lily is, of course, the *Lilium candidum.*

We now come to the New Testament lilies, as in Matthew vi. 28 and Luke xii. 27, and here there are obviously differences of opinion. Some think that the word here is much more a general term, meaning beautiful wild flower. Others say it was probably a gladiolus or an iris or even a Martigon lily. Anyway, the point of our Lord's discourse was the three words "How they grow". It does not matter very much whether the word lilies here was a generic term for some beautiful wild flowers, or meant a long spray of Gladioli. One has only to examine any flower to see how it grows, to realize how wonderful is God's nature, for we soon pass on to the even more authoritative words, "Yet, I say unto you".

However, there are experts who insist that this lily of the field refers to the *Anemone coronaria.* The point here is that these anemones grow wild, are scarlet-purple in colour, and when in a large drift they certainly surpass "Solomon in all his glory". No one quite knows when these anemones arrived in Great Britain, but it was probably during the time of the Crusades, though the official date given by experts is 1596.

Some have wondered whether in Song of Solomon ii. 1, 2, the lily among the thorns is quite different from the lily of the valleys. One can say, almost authoritatively, that it is not the Lily of the Valley that we know today, and as our Lord seems to have connected the glorious scarlet-purple anemone with Solomon, there is some basis for believing that it is this anemone which is called the lily of the valley. One could then have the tall Madonna lily growing among the thorns, and the great

LILY. (Lilium candidum).

sweep of the bright anemones in the more fertile part of the valleys.

There is justification for this thought, because in the Song of Solomon vi. 3 we read about the beautiful gazelle feeding among the lilies. Now a gazelle would feed in the valley in which the best of the soil had been deposited over the years and so the pastures were particularly inviting. The lowest field at the Thaxted Horticultural College, when I was Principal there, had a depth of soil of forty inches, whereas the other fields that sloped down to it had only eight or nine inches of soil. This shows the great advantage of the lower reaches of a farm, where the soil is particularly good.

What actually happened when "my beloved went down into his garden" (Song of Solomon vi. 2) is difficult to conjecture. Did he actually pick Madonna lilies for putting into bowls in the house, or as he went down there to feed was it that he was digging up the lily bulbs, which in the Orient are eaten? If this is so, then it is probably the Tiger lily to which this refers, but there is no proof that this grew in Palestine at the time. There is no question that the word "feed" is properly translated. It is the word "*rah*" which is used in Genesis xxxvi. 24, "he fed the asses", or 1 Samuel xvii. 15, "to feed his father's sheep" or even in Hosea iv. 16, "the Lord will feed them as a lamb".

The word *shoshan,* meaning lily, is undoubtedly connected with Shushan, the palace in which Ahasuerus lived (Esther i. 2). It may have been "the palace of lilies" for in those days water lilies were much beloved—and reverenced, for that matter—the Nymphaea lotus being the lotus flower, a type of water lily. It may be said, therefore, that this particular lily, about which there are

LILIES OF THE FIELD. (ANEMONE CORONARIA).

many mystical stories, was planted in pools made in gardens around the palace, or even, as is more probable, within the palace courtyards.

Tradition has it that when Eve left the Garden of Eden she shed real tears of repentance and from those tears there sprang up Arum lilies, the spiritual application being, of course, that true repentance is the beginning of beauty. Again, tradition has it that the Arum lily should be called the Gethsemane lily, because the plants were found growing in the garden after our Lord's agony, and where the drops ofn sweat fell these beautiful lilies grew.

In Song of Solomon ii. 1. Lilies of the Valley are mentioned. I think this plant is the *Hyacinthus orientalis*, which grows quite commonly in Palestine. The flowers are a deep blue and fragrant and in the Spring around the Lake of Galilee they are most attractive.

Another lily I found there was the *Lilium chalcedonicum*, and I have often wondered whether our Lord could have been referring to this lily when talking about King Solomon. But when you turn to Matthew vi. 30 and read, "which today is and tomorrow is cast into the oven", one wonders whether the plant was not a lily at all but the *Chamomile anthemis palaestina*. This is made into hay. It has scented aromatic leaves and white, downy flowers. It is in flower at haymaking time and so would very likely have been growing around and about our Lord's feet as He was speaking. It certainly is very common in Palestine.

Onycha

God said to Moses in Exodus xxx. 34 "Take unto thee sweet spices, stacte, and onycha and galbanum". These sweet spices, as they were described, were to be mixed with pure frankincense in equal parts and they were to be made into a perfume by the apothecaries. It was to be a "holy" perfume and was to be used in the Tabernacle only. In fact God said that if others attempted to make their special scent they should be "cut off from his people" which means, of course, excommunicated from the church.

Now, what is this onycha? We get the clue, perhaps, from the Arabic translation of the Bible, where the word "onycha" disappears and "ladana" is introduced instead. "Ladana" refers to the Rock Rose (*Cistus ladanifer* or *Cistus salvifolius*). It is this beautiful flowering plant which produces a scented gum resin and might therefore easily be the substance which Moses was instructed to use. The Rock Rose produces flowers about three inches across, the petals are white, but they have a lovely scarlet-rose blotch at the base. The hundreds of stamens in the centre of each flower are golden in colour, and this makes the blooms most attractive. Incidentally, I am told that the Greek word for "fingernail" is "onycha" and it is thought that the markings in the centre of the petals may have given rise to this name for they have a "finger-nail" look about them.

Of course, it may well be that the onycha is some perfume that the Children of Israel took with them out of Egypt when they "spoiled the Egyptians". Moses would know well that most of the women had "stacte, onycha and galbanum" hidden somewhere, and these were to be used for the worship in the Tabernacle. Anyway it does seem that the use of these special perfumes disappeared when the Children of Israel arrived in the Holy Land, and maybe this means that they were quite unprocurable in the land of Canaan.

Poppy

Most people consider that in Matthew xxvii. 34—"They gave him (the Lord Jesus) vinegar to drink mingled with gall"—the word "gall" here refers to the plant *Citrullus colocynthis.* This plant produces a round cucumber-like fruit more like the apple-cucumber, but the colour of an orange. The pulp is poisonous, bitter, and I am told is sometimes used as a purgative. It is common in the Mediterranean region. The fruits when dried and powdered can be used for keeping moths out of wool.

It has been suggested that the word "gall" in this text of Matthew refers to opium and that the vinegar offered to our Lord contained the juice of the Opium Poppy *(Papaver somniferum),* which, had He accepted it would have sent Him into a heavy sleep. If this were so, then this may well be the reason why our Lord refused it, because it was part of His divine purpose to suffer.

It is unlikely, however, that the Roman soldiers would have wanted to mitigate His suffering by offering Him a narcotic. It is therefore more likely that the word "gall" used here refers to something bitter added to the vinegar and some people suggest, therefore, that it was really myrrh. I would like to know whether the Roman soldiers in those days did drink a kind of vinegary wine with a dash of myrrh in it, very much as people take alcohol today with a dash of what is called "bitters". If this is so, then the soldiers just offered him what they were drinking themselves.

On the other hand, there is little doubt at all that it could have been the Poppy *(Papaver somniferum)* for this grew quite happily in the Egyptian gardens around the Nile, as we know from ancient records. The flower of the opium poppy is very beautiful indeed, being either pure white or lavender in colour. The petals have a beautiful stain of purple at their base, while the leaves are a pretty bluey-silvery-green. If one makes a little incision in the poppy head three days after the petals have fallen, a milk-like juice will exude and this will solidify at the end of a day. These solid drops could have been stirred into the vinegary wine, the result being an opiate.

Rose

It really is an extraordinary thing that the Hebrew word "*chabatstseleth*" in Isaiah xxxv. 1 should be translated "rose". It is no more rose than the Rose of Sharon in Song of Solomon ii. 1. It is probably the *Narcissus tazetta*, though some consider that it may be the *Crocus sativus* or Saffron crocus. This is, of course, the plant that gave the name to Saffron Walden which was only seven miles from the Thaxted Missionary Horticultural College. It also gives the name to the yellow saffron powder which is used principally by Cornish women to make their delicious cakes. The British Saffron was much in demand, but it must be remembered that in order to produce only a quarter of an ounce of the dry saffron powder, one needed one thousand stigmas from this particular crocus plant.

However, there is in my mind little doubt that the saffron referred to in the Song of Solomon iv. 14, the Hebrew word "*karkom*", is the *Crocus sativus* (which will be dealt with under the next heading), whereas the Rose of Sharon is the Narcissus.

There were undoubtedly wild roses in the country of Palestine, but the flowers were small and they probably would not last in bloom more than a day. The translators of the Authorized Version obviously had in mind

SAFFRON (CROCUS SATIVUS)

some very beautiful plant which would give a great splash of colour and would be in complete contrast to the desert and, of course, in common parlance today we still talk of the desert blooming as the rose, using quite literally the text of Isaiah xxxv. 1.

It is interesting to record that Farrer Fenton in his translation of Isaiah xxxv. 1, 2 has it, "Let the moorland and desert rejoice and laugh and with lilies spring up—be fruitful in produce and laugh, ah, yes! laugh and shout."

If we turn to the *Tazetta narcissi* in any good catalogue, we shall find them described as perfectly hardy and they may be left undisturbed in the garden for years. Actually there are many different varieties of Tazetta today, such as Martha Washington, Primrose Beauty, Scarlet Gem and St. Agnes. They are what is described as the bunch-flowered narcissi and have small, coloured cups, with snow-white petals around. There are usually five flowers to the stem.

The prophet had in mind the change that took place from an apparently dry, wide terrain, which looked like a desert, to the thousands of narcissus-like leaves which shot up through the ground the moment there was a rainfall. The leaves were followed "hot-foot" by the flowers and so it seemed that in almost a few hours you had the striking change from bare soil to myriads of beautiful blooms.

Only those, like the writer, who have lived in South Africa know what tremendous changes there can be almost overnight. The veld seems dull and lifeless and desert-like but after the parched soil receives its first good rainfall, millions of little plants leap into life and flower profusely—flowers like the Star of the Veld, which can be grown in British gardens today.

In the second book of Esdras ii. 19 it states, "Whereupon there grew roses and lilies." Most Bible students agree that this probably refers to *Rosa phoenicia.* This is a wild rose commonly called the Phoenician, a bush which grows eight or nine feet high. It produces masses of scented, single white flowers, centred with golden stamens. It has been found growing at heights of five thousand feet and therefore it can be regarded as being suitable for "the seven mighty mountains" which are mentioned in the second book of Esdras.

Saffron

Saffron today is the prepared stigma of the *Crocus sativus.* Its root is perennial and consists of a solid bulb. It blooms in Britain in September and October and was probably introduced to this country from Asia or Palestine hundreds of years ago. The flowers are picked early in the morning just as they begin to open and the pistils are carefully picked out, for they are the part containing the Saffron—the rest of the flower being useless. In olden days they used to dry them on a portable kiln, with a good deal of fire heat to evaporate the moisture.

The Greeks and the Romans made use of Saffron for perfuming their rooms and theatres, as well as for seasoning their dishes and scenting their salves. In the days of our grandmothers, a cake saffron was often adulterated with mixtures of the fibres of the petals of the golden thistle and common marigold, and the test was to infuse a very small quantity in some hot water, when the true saffron became easily distinguishable from the other ingredients. It is said that the average produce per acre of dried saffron in Saffron Walden was only two pounds the first year, but twenty-five pounds an acre after that.

Actually the origin of this *Crocus sativus* seems to be lost in the mists of antiquity. It has certainly been known for some four thousand years and it is included among the spices and flavourings Solomon saw used in his kitchens in the Song of Solomon iv. 14. Just as the old saying goes, "A woman, a dog, and a walnut tree; the more they are beaten, the better they be", so the great Roman writer Pliny states that the saffron bulb should be beaten or well trodden down. I have always imagined that this must be because the firming of soil would help to encourage the production of "bulblets", and that was obviously the main way in which these valuable bulbs were propagated in those days.

Star of Bethlehem

When I was young I was always tremendously worried about the verse in 2 Kings vi. 25, "and there was a great famine in Samaria: and behold . . . an ass's head was sold for four-score pieces of silver, and the fourth part of a cab of dove's dung for five pieces of silver." It always sounded to me so extraordinary that in a famine people should eat dung. Later on, I discovered from a concordance that the Hebrew name was *dibyonim* which can be translated "a roasted chick pea". Linnaeus, the world-famous botanist, believed some two hundred years ago that this term "dove's dung" referred to the bulbs of the plant Star of Bethlehem, *Ornithogalum umbellatum.* It is said that Cleopatra knew of this bulb, which when dried and ground was used as a kind of food or medicine. Parkinson, the British herbalist of the 1800s, described it as "sweeter than a chestnut", presumably referring to roasted or boiled chestnuts, for the bulbs of this *Ornithogalum* are said to be poisonous when eaten raw.

STAR OF BETHLEHEM. (ORNITHOGALUM UMBELLATUM)

Certainly this bulb was known in Palestine. It was available there at the time of the siege of Samaria and it can easily be that a certain quantity of these baked bulbs were sold for the five pieces of silver.

Flowers as a whole

It is curious that there should be so little about flowers in the Word of God, but the people of Biblical times lived simple lives, and furniture would have been meagre and there would be no occasional tables available for bowls of blossoms! The gardens themselves were largely groves of trees, where people could sit in the shade. There were other flowers not mentioned in this chapter, for they are those concerned with fruit trees.

CHAPTER 4

THE FRUIT TREES IN THE BIBLE

IT IS DIFFICULT to be sure which is the most important fruit tree of the Bible. Some would claim the vine. This certainly is mentioned again and again, right from the time of Noah (Genesis ix. 20) and all the way through to Revelation xiv. 19, where there is a description of the angel gathering the vines of the earth and casting them into the great winepress of the wrath of God. Others might argue that the olive is extremely important. It was cultivated from time immemorial, for it was an olive leaf that was plucked by the dove and brought back to Noah in Genesis viii. 11. Olives in fact are mentioned very many times throughout Holy Scripture, the last reference being the two olive trees, "standing before the God of the earth" in Revelation xi. 4. The third claimant might be the fig, which is mentioned in Genesis, "they sewed fig leaves together", Chapter iii. 7, with the final reference in Revelation vi. 13, "as a fig tree casteth her untimely figs".

Tree of Knowledge

Perhaps no fruit tree has been more illustrated than the Tree of Knowledge in Genesis iii. 6. Artists of all countries have pictured in many differing ways the picking of the fruit by Eve and the giving of it to her husband. In the picture somewhere is usually the Devil, depicted as a serpent. Actually he was not turned into a crawling serpent until after the temptation had taken place, when the Lord God cursed him. However, we are not going to conjecture about the temptation; the fact is that it did take place and that Adam and Eve fell. All down the ages, people have wondered as to the fruit itself. Some have claimed that it must have been the pomegranate, for this is a very attractive fruit. On the other hand, though oranges are never mentioned in the Old Testament, or in the New Testament for that matter, it has been said that the so-called apple was undoubtedly an orange. The argument is based on Proverbs xxv. 11, as "apples of gold in pictures of silver", the tree of knowledge would indeed have been exquisitely beautiful with its golden apples glittering among the silvery white flower. The wonder of the orange tree is that it produces its leaves, fruits and flowers simultaneously; and because it is thus so vigorous it is called by the Eastern people—a tree of life.

In the Apocrypha, the apple is described as having a refreshing perfume, with the fruit really sweet and with the foliage affording ample shade. This description is far more true of an orange tree than it is of an apple, and especially one growing in that part of the world. It is not that the apple did not grow in Palestine, but the varieties grown there could never in any circumstances be regarded as refreshing or sweet. Things are changing in Palestine today and we may have delicious apples grown there before long.

I have never been able to discover why some people have claimed that the tree of knowledge was the quince, unless it is because it is golden in colour, and is shaped somewhat like an apple. It is certainly unlikely to have been the lemon, for this is not sweet, but there is little doubt that King Solomon grew both limes and lemons, which would have been introduced to him by the Persians. He is said to have planted "trees of all kind of fruits" (Ecclesiastes ii. 5) and it may be that his ships, which undoubtedly traded with India, brought back with them various rare fruit trees from that country.

The last claimant to the title of the tree of good and evil is the apricot, and those who take this line argue that when Henry VIII first ate apricots, they were called Armenian apples; and that great Palestinian traveller, Canon Tristram, thought that the apple must be the apricot, because it met the requirements of the word pictures painted in scripture. The flower of the apricot appears long before the fruit. It does not give quite the same shade as the orange and I find it difficult to understand why it was ever suggested. However, the fruit of that particular tree might not have been like any that is known or grown today.

Of course, the word "fruit" in the Bible is sometimes used figuratively. For instance we read "The land shall yield her fruits" in Leviticus 25:19, or "Blessed shall be the fruit of thy body" in Deuteronomy xxviii. 4. While in Proverbs xi. 30 it talks about the "fruits of the righteous."

It is wondered whether the almond ought to be included under the heading of fruits because it is the seed in the centre of a fruit. It could be argued also that "balm" which is mentioned in Jeremiah viii. 22. may come from the fruit of the Balantine tree. The cucumber is a fruit, but it is classified as a vegetable and is therefore found in that chapter, while the mandrake found in Genesis xxx. 14 has been called Devil's apple", for the yellow fruits are like apples.

I often wonder whether the corrupt fruits mentioned in Luke iii. 9 are really trees that have arisen from self-sown seeds and which might therefore bear useless fruits. The "summer fruits" in Amos viii. 1 seem to refer to the fruits that ripen at the end of the season, while the "hasty fruits" mentioned in Isaiah xxviii. 4 are really the early ripeners of the summer which must be eaten before they start to rot.

Apple

There are many conjectures as to the references to apples in the Bible. For instance in Proverbs xxv. 11, Song of Solomon ii. 5 and in Song of Solomon vii. 8 the Hebrew word *tappuach* appears to be quince, as for instance it does in Joel i. 12. Whatever the tree was, it certainly afforded good shade, and its fruits were not only enticing, but also sweet to eat. Furthermore they were fragrant, and golden, with (as some people consider) silvery leaves, though the author is sure that the silveriness refers to the flower petals.

The apple certainly could have grown in Palestine, but does not seem to fit into the picture of the description given in the Bible. The fruit is much more likely to have been the golden apple, that is the orange.

Those writers who have criticized the suggestion that the apple might be an orange, always said that an orange tree could not be large enough for it to produce much shadow to sit under, as is mentioned for instance in Song of Solomon ii. 3. This, of course, is untrue; on the other hand, Dr. W. N. Thomson reported in 1886 that wonderful apples were grown at Askelon, and stated that these had both the right smell and colour to fit into the description given in Proverbs xxv. 11 and in other places.

In Cyprus I found that the apricots were called "golden apples"—could this fruit therefore be the answer to our problem? The Cyprus apricot grows to a height of thirty feet and gives good shade, the fruits are golden and perfumed, the leaves are of a pale colour, the undersurfaces are covered with down. The flowers are white with a pink tinge.

There is also the reference in Deuteronomy xxxii. 10, Psalm xvii. 8, Proverbs vii. 2, and Lamentations ii. 18 to "Apple of the Eye"—the Hebrew word *Ishon* meaning "bath". Thus the translation should truly be "bath of the eye". In Zechariah ii. 8 the Hebrew word used is *babah*. This means "gate of the eye".

Brambles

Though we today like to go out into the hedgerows to gather blackberries, the farmer who allows these brambles to spread into his fields finds that they quickly ruin his pasture, and to him they are weeds. Most writers, therefore, include the brambles of the Bible under the section of pernicious weeds. The thorns referred to in Matthew xiii. 7 by our Lord, may have been thistles, or brambles.

Unfortunately, there are many different Hebrew words that have been translated "bramble" or "briar", or even "thorn". The Hebrew word *choch* of Isaiah xxxiv. 13 which has been translated "bramble", is undoubtedly a thistle. There were probably two brambles growing in Palestine in the time of our Lord. *Rubus sanclus*, the indigenous bramble, and *Rubus ulmifolius*, the elm leaf bramble. I found them both to be evergreen with prickly stems. Both of them spread by means of underground stems and by suckering—somewhat like raspberries. These plants grow in thickets and are happy near water.

It is in Luke vi. 44 that we obviously find the fruiting bramble, for our Lord suggests that you would not be gathering grapes from a blackberry bush. It is also thought that it is the fruiting blackberry that Gideon referred to in Judges viii. 7 when he was going to whip the enemy with the long blackberry canes.

The bramble in the parable found in Judges ix. 14 is the Hebrew word *atad* and is probably the buckthorn, while the brambles mentioned in Isaiah xxxiv. 13 are undoubtedly thistles.

Fig

The first mention of the fig tree is found in Genesis iii. 7 when Adam and Eve tried to cover up their own sin by sewing fig leaves together and making "aprons". I much prefer the translation in Moffatt's rendering which makes the sewing of the leaves together into a girdle. This is much better than the translation of the Geneva Bible which refers to breeches. This, of course, is a picture of what is found later in Isaiah lxiv. 6, "all our righteousnesses are as filthy rags".

We find figs used as a remedy or as a poultice, as in 2 Kings xx. 7. This is repeated in Isaiah xxxviii. 21, when the story of the healing of the boil is emphasized. Personally, I like the use of the word "naughty" as in Jeremiah xxiv. 2 when the various types are described, the very good ones, the first ripe ones, and the "naughty" ones. This, of course, means the bad figs. The Hebrew word *pag* in the Song of Solomon ii. 13, is the small unripened fig which remains on the tree throughout the winter.

The word *bickurah* in Hosea ix. 10 is the early fig, about which more anon, and the term *debelah* of 1 Samuel xxv. 18 is the cake of dried figs which is, of course, the way in which figs are stored in the winter, which again and again were used for food. These cakes of dried figs are similar to those that are often eaten during the Christmas festivities in this country. Such figs and, of course, the modern "Syrup of Figs" are used as an answer to constipation.

The fig was so important to the Jews that every man liked to have his own fig tree as well as his own vine (1 Kings iv. 25). A well-cultivated tree should produce fruit for ten months of the year. In this country, minute figlets on the young wood are invariably killed during the winter, and so you seldom get the earliest crop; and because the frosts come too early (usually at the end of September) you never get the October and November figs ripening as they should.

Fig trees may even live as long as four hundred years. They will reach an immense size, having a beautifully thick trunk and smooth bark. The fruits are pear-like and are delicious to eat. They contain the imperfect flower right in the centre of the fruits. There are some rounder varieties of fruits, but in all cases a small hole is found at the farthest end from the stalk, through which a burrowing insect may enter. This insect is known as a fig wasp, and fruits never set properly until this creature has done its fertilizing work. The gritty seeds are the true fruits. The fig that we normally refer to as a fruit is really the receptacle that holds these miniature seed-like fruits.

In this country we had figs growing prior to the Christian era and Pliny describes six varieties. The trees may have got lost, or perhaps, were destroyed by the ancient Britons, because history tells of the re-introduction of the fig by Thomas-a-Becket and the planting of the first fig tree at Tarring in Sussex.

We must now refer to the cursing of the fig tree by our Lord in Mark xi. 13 and in Matthew xxi. 19. It was the time of the Passover, in the month of April—some have therefore claimed that you could not have expected any figs to be on the tree at that particular time of the year. Fresh young leaves, yes, but the fruit would not be ready! If, on the other hand, the tree concerned was one of the very early varieties known today in Palestine, bearing large green fruits, then it should have been possible to have had ripe fruits at Easter in a warm sheltered valley.

It must be remembered that the fruit often starts to be ready before the leaves and this is especially true in the case of an early variety. Anyway, our Lord who made the trees would certainly have known whether this particular type should have been fruiting or not. He certainly would not have been looking for the fruits long before their proper time, as some critics have tried to suggest.

Further critics have suggested that our Lord had no right to attempt to pick the figs, because they were not his, but our Lord knew the laws for he had made them, and Moses had said that fruits on trees by the wayside could be plucked by the passers-by. Josephus, the great Jewish historian, makes this quite clear in his writings. Even today, I am told, in Palestine fig trees by the side of the road are, so to speak, common property and travellers can pick a fig or two as they pass by. The whole point of this "parable" was to do with the text, "by their fruits, ye shall know them" (Matthew vii. 20). There are so many people who make a profession of Christianity, with the result that their "leaves of profession" are useless and they are "nigh unto cursing".

In Luke xiii. 8 we see quite clearly the need for cultivating the soil around fig trees and for giving them the necessary farmyard manure or compost. This seems to be particularly true in the case of young trees. Palestinian experts tell me that in the drier areas fig trees could certainly prove to be quite useless at the end of three years if the ground around them was not cultivated or mulched with dung. Good, well-flavoured, luscious figs only result where figs are properly cultivated. Fruitfulness, therefore, depends on careful culture. There is a spiritual parable there.

Locust Bean

Readers who have travelled on holiday to the south of Spain or to Majorca, or to other parts of the Mediterranean will know of the Locust Bean tree, *Ceratonia siliqua*. The beans often hang on the trees long after the leaves have fallen. They are shaped somewhat like a broad bean, only they are a dark reddish brown, and they are slightly curved. Children sometimes hold up two of these dry beans (for they dry in the sun quickly, even on the trees) on either side of their foreheads, and make them look like goats' horns! Now in Luke xv. 16, in connection with the story of the prodigal son, the Greek word for "husks" really means "little horns".

As it is known that these pods are fed to pigs, and other animals, it is thought that these locust beans, which are the fruits of the Carab or Locust tree, must be those to which reference is made. I have eaten these dry beans, and they are not unpleasant to the taste; in fact they are sweet. In parts of the Mediterranean they make not only syrup from them, but wine too.

John the Baptist in Matthew iii. 4 and Mark i. 6 is described as eating locusts and wild honey, and some writers have suggested that this refers to the Locust Bean and not actually to the locusts or large grasshoppers. In some parts of Europe the Locust Bean is known as "St. John's Bread". There is no doubt, however, in the author's mind, that the text actually refers to the grasshopper, for once the wings and legs have been pulled off the body, the creature is regarded (even today) in the East as something delicious.

Melon

Melons are only mentioned in Numbers xi. 5, where the Hebrew word *abattichim* is used. These are probably water melons, which are still grown today in many of the villages in Palestine. It is one of the wonders of the vegetable kingdom that such melons, which are beautifully green outside and wonderfully red inside, should be able to grow in such dry soil. The reason is, of course, that the roots penetrate deeply downwards through the dry sand and reach some moisture well below. In addition, there is no doubt that the large leaves absorb moisture from the heavy dews which fall during the night.

It is no wonder that these large, beautifully cool, juicy melons should have been remembered by the Israelites in the wilderness. The Arabic name *butteekh* appears to be only a variation of the Hebrew. There are some who consider, however, that the word "melon" is really the sweet melon, and maybe Cantaloupe Melon, and that the cucumbers mentioned may have been, in fact, water melons.

Olive

Olives and olive trees are mentioned fifty-seven times in the Bible, that is if one discounts the Mount of Olives and the "Olivet" which described a ridge of hills on the east of Jerusalem, about a mile from north to south. It is obvious that the Olive is one of the oldest known trees, for we read of it in Genesis viii. 11, "in her mouth was an olive leaf pluckt off".

It is supposed that both the dove and the olive leaf became emblems of peace, as a result of this incident. Ever since people have talked about "holding out the olive branch" and Picasso amongst others has made famous the dove of peace. Doves love to build their nests in olive trees in Palestine, even today! Those who have argued that the olive does not flourish on the tops of mountains, and at any rate not in the Ararat area, should remember that the dove had all day to make its long flight, because it did not return before evening. It probably, therefore, flew a great distance to obtain its olive leaf.

The olive was one of the blessings of the Promised Land and is certainly abundant at this time in Palestine. It will grow on the mountainside where there does not seem to be much soil and it will crop very heavily, producing fruits from which olive oil is extracted. The wood of the trees is useful because it is finely grained, and turns to a nice rich amber colour after being cut and planed. They chose olive wood for the posts and the doors of the temple, as well as for the Cherubim.

Writers like Pliny and Loudon seem to infer that olive trees can live to be nearly two thousand years of age. At any rate, it is known that even when trees have been cut down, suckers invariably grow up and so in a few years what appears to be a ruined olive orchard can easily be a heavy cropping acreage again.

In Hosea xiv. 6 the prophet says; "his beauty shall be as the olive tree". Now, beauty, we are told, is in the eye of the beholder, and to the outsider it may be said that there is nothing beautiful about an olive tree. Its trunk may be twisted, its leaves small, its branches uneven and pendulous, and the berries of such a colour that they are almost insignificant. To the Spaniard, however, who knows that a heavy crop is going to make all the difference to his livelihood for a year, a well-laden olive tree is very beautiful indeed!

Furthermore, an olive orchard in the southern part of Spain, and in Palestine too, for that matter, provides the most beautiful shade and protects from the heat of the noonday sun. The traveller, or the farm labourer, looks upon the olive tree as very beautiful when it provides him with the shade he needs. It is from this point of view that I think one must consider the text in Hosea.

There is so much that can be said about the olive in Scripture that one hardly knows where to begin or end. For instance, there is the wonderful provision for the fatherless, the widow and the stranger found in Deuteronomy xxiv. 20 and in Isaiah xvii. 6. The Hebrew farmers, when harvesting their olives by beating the branches were not to attempt to beat down the very last berry. A few olives were to be left on the topmost boughs for those who were too poor to grow their own. They were to be allowed to come and glean.

It has always surprised the writer to discover the amount of olive oil that can be got from one tree. First of all you get two crops a year as a rule, and then from an old tree you can easily pick sufficient olives to produce twenty gallons of oil. One cannot over emphasize the importance of this crop in a rocky arid district. There may be no grass for the cow to graze. There is probably little on which even the goat may live, and so the peasant in Palestine and in parts of Spain will use olive oil instead of butter. King Solomon, for instance, had to send olive oil to the foresters of King Hiram for their use.

Oil is the symbol of the Holy Spirit, and we read about

this clearly in the Parable of the Foolish Virgins. It was refined so that it could be the perfect oil for putting in the lamp which was lit before the altar in the Tabernacle. It provided the holy ointment with which the vessels of the Lord were anointed.

As the trees grow in such rocky land, it is no wonder that Moses refers to "oil out of the flinty rock", in Deuteronomy xxxii. 13. It may be that he had in mind the circular stone troughs into which the gathered olives were placed, so that they could be rolled by those exceedingly heavy stone wheels. In parts of France these wheels and troughs are still used for crushing the cider apples. The result of the pulping carried out by the wheel would be the production of the oil and so much of it perhaps that in Job xxix. 6 we read of "rivers of oil".

A large number of varieties of olive are cultivated, and these are propagated by being grafted on to the seedlings of the wild olive—*Olea europaea.* There is what may be called a wilder form *Olea oleaster,* which produces four-angled shoots which are quite thorny. The small fruits it bears are roundish, and inedible. It may be this fruit to which St Paul refers in Romans xi. 17, 18, and 24. The problem of the passage is, however, that the wild olive is grafted onto the good variety, and though it is known that the root part of the tree (as in apples in this country) does have a lasting effect upon the graft, yet one could not believe that a wild useless variety would be converted into a good fruiting kind just by grafting. Theologians therefore say that in the spiritual realm this can and in fact does happen, and that God has the omniscient power to do things that are contrary to nature. We Gentiles, therefore, the wild race, can be grafted into the good olive tree of "The

Church" and we can indeed bring forth the type of fruit that will lead to eternal life.

Olive oil is often used in the Bible in cases of sickness. The apostles, for instance, anointed sick folk with oil and caused them to be healed, as in Mark vi. 13, and some of us have seen men and women miraculously raised up from beds of sickness as a result of the special anointing by the elders as detailed so clearly in James v. 14. The poor traveller in the story of the Good Samaritan had oil poured into his wounds (Luke x. 34), and even today in the rural parts of Spain olive oil is used in this way, to help heal wounds.

In the description of our olive tree we have failed to make reference to the beautiful little white flowers it produces in abundance. It is claimed by botanists that only one flower in every hundred actually produces a fruit, and thus from a large olive tree the petals will fall in abundance, just like "snow in summer". Eliphaz mentions them in Job xv. 33, when he warns men not to trust in vanity which melts away like the thousands of useless petals that seem to be contemptuously thrown off by the large olive tree, as being nothing to her at all. She concentrates on the blossoms that have set and her "fatness" feeds them.

One cannot over emphasize the importance of the olive to those who lived in the Bible days. It was daily food, with a crust of bread, for the farm labourer. (In fact even in 1948 it was the normal breakfast of the Superintendent of the International Mission to Miners in Spain because food was so scarce.) Travellers took with them twenty or thirty olives, and the paper-like loaves which the boy gladly gave to our Lord. Most of the meals would be cooked in oil. The lamp that was lit to provide light in the evening was fed with oil. The soap was made from olive oil and in fact there are even today special soaps that, it is claimed, are made from olive oil only.

It is most important to realize these things when you consider Habakkuk iii. 17, 18, for even if the olive crop failed, yet the writer would rejoice in the Lord. This seemed almost the worst calamity that could come to him, to have no olives at all to eat, to cook his food with, to wash his clothes with, to use medically and of course he would have no light at night-time either.

In Spain the olive trees bear fruit in the winter, and they are picked at the end of December, January, and sometimes right into February if there is a large or late crop. Unless the trees suffer from drought or blight they bear every year. The usual method of gathering is for tarpaulins to be spread around the tree; the men then beat the branches with canes, and the women on their knees pick up the olives and put them into baskets.

The Bible speaks of the olive trees being shaken, see Isaiah xvii. 6 and xxiv. 13 and when the olives are ripe it is sufficient to shake the branches to bring down the fruit. In Spain they use long poles to beat the uppermost branches because it is difficult to shake a branch that is so high up. See also Deuteronomy xxiv. 20. The actual harvesting of the olive may have to be done at a particular time, and the work will proceed from early in the morning to late at night.

Very often the olives are planted on the sides of mountains, and so even after the fruit was gathered it had to be carried quite long distances to the farm. The olive groves are seldom enclosed, and so the trees of one man might be next to the trees of another. This also makes it important to do the harvesting round about the same time. It is no wonder that you find olive trees along the Mediterranean coast, for they like the sea air, and presumably benefit from the mists that arise as a result of the trees being near the water. Moses causes the Levites to make this quite clear in Deuteronomy xxviii. 40, for the people were to have olive trees along all the coasts.

In likening the olive to the Jewish people, God seems to make it quite clear that their purpose was to produce fruit. They were to bring results. They were not to be vain, nor should they want "kings to reign over them". Like the olive, they were to last, they were to give shade to others, they were to be quietly cropping, even though not as beautiful as the tall cedar or as graceful as the date palm. They were to yield peaceable fruits in places where other trees could not grow. They were to have a glorious spiritual role in a world that knew only pomp and power. Hosea emphasizes this point in chapter xiv. 6, while Jeremiah puts the seal on the matter by saying "The Lord called thy name a fair green olive tree" in Jeremiah xi. 16.

Until I saw an olive grove, I had really no idea what David meant when he said in Psalm cxxviii. 3: "Thy children shall be like little olive trees round about thy table." It is really a picture of the young olive trees growing around the base—usually some distance away—of a big parent olive tree. When, as it were, "Papa" gets too old to bear sufficient fruit, younger trees arise, which start in themselves to crop, and so help to bear the burden of the parent. They even appear to protect the parent tree from the cold winds and winter snows, and thus they are the picture of how children should indeed honour their parents, not only when they are young, but when they are old also. It also demonstrates how they themselves, by starting to be productive, both in a temporal manner and spiritually, can start to help those who are getting aged and can also add to the wealth and spiritual welfare of those round about them.

As far as I can discover, the only "Oil tree" mentioned in the Bible is found in Isaiah xli. 19: "I will plant in the wilderness the cedar, the shittah tree, and the myrtle, and the oil tree". The word used there is *shemen* and I find from my concordance that this is exactly the same word as is used in 1 Kings vi. 23, 31, 32, and 33. In these four references the translators have used the word "olive tree" and not "oil tree". Following on this line of thought I turned further on in the commentary, to find *shemen* again mentioned in Nehemiah viii. 15: "Go forth unto the mount, and fetch olive branches". Now once again we get the word *shemen* used, but this time to describe pine branches, for the Hebrew word for olive branches is *zayith*, which is the same word used throughout most of the Old Testament to describe the olive tree or olive branches.

If these two words are to be translated "olive branches", then it would mean that in Nehemiah viii. 15 the word "olive" is mentioned twice. I cannot, therefore, help believing that the word *shemen* cannot properly be translated olive tree and the answer may either be that the Hebrew writer was trying to make the difference between a cultivated and a wild olive, or that the second word *shemen* is really the *Eleagnus*, which comes from "elaia" meaning olive and "ognos" which was the Greek name of *Vitex agnus-castus*. This *Eleagnus*, sometimes called the Oleaster, has fruits of a pleasant flavour, half an inch in length, with stalks up to an inch long but it is not an olive.

Palm

It is said that there are two hundred and fifty different kinds of palms, but there is no doubt that the varieties which bore the very pleasant fruits were the ones which were popular in Palestine. Normally speaking, the Hebrew word used for palm tree is *Tamar*, but the special palm tree under which Deborah dwelt in Judges iv. 5 is *Tomer*. Jeremiah in fact delights in emphasizing that the heathen are like this special type of palm because he says in Jeremiah x. 5, "They are upright as the palm tree, but speak not". Whether this, "speak not" means that they do not bear fruit, I would not know. It may quite easily be that Deborah's palm tree was not a fruiting type and perhaps she dwelt under it—only because it produced an abundance of foliage, for often trees that do not fruit make more wood.

One gets the feeling from the Hebrew that the word *tamar* definitely describes a fruiting tree. Read, for instance, Joel i. 12 where he says, "The vine is dried up and the fig tree languisheth, the pomegranate tree, the palm tree also . . . are withered: because joy is withered away from the sons of men." All the trees mentioned are fruiting trees. David, too, talks about the righteous flourishing like the palm tree, and he goes on to say, "They shall bring forth fruit in old age;" Psalm xcii. 12-14.

Solomon on the other hand uses the palm tree to describe height and slenderness and beauty. The Arab poets describe beautiful women as palm trees. Thus in the Song of Solomon we read in vii. 6, 7 "How fair and how pleasant art thou, O love . . . This thy stature is like to a palm tree . . ." "Tamar" often referred to a tall willowy girl.

Palms are of tremendous importance in the East, and it is claimed by Arabs and Jews alike that almost all the material wants of man can be furnished by this "family". It produces the fibre for ropes, the material for thatching roofs and for making baskets, wax and oil, nuts and fruits, and liquors by fermentation. This perhaps is the reason why it is the symbol on the earliest coin. Phoenicia got its name from the ancient Greek name of the date palm, *Phoenix.* It is the *Phoenix dactylifera,* the fruit of which are the common dates, and are the only ones used for food. The trees can grow to a height of one hundred feet and to ensure that the female blooms set properly, the Arabs often cut off the male inflorescences and hang them in the female trees. Many is the child who has sown a date stone out of doors, and has seen a plant grow. Unfortunately, however, the date palm is killed by the frosts of this country, and therefore the little plants must be grown under glass.

Because the date wine or date brandy is so strong, it is thought that the text in Proverbs xx. 1 "Wine is a mocker, strong drink is raging, and whosoever is deceived thereby is not wise," refers to this particularly potent drink. The fruit does not ripen all at one time, and so one can pick the dates over quite a long period. One needs to be a skilled climber to do the date harvesting, and for this reason many peasants find it more convenient just to shake the trees and to catch the dates on sacking held below.

Those of us who have eaten dates at Christmas know that these are quite pleasant, but it is not sufficiently realized that the ones we get in Great Britain are sun-dried. A tree has to grow a great many years before it starts to crop, and some Middle East experts have said that the date palm does not really reach maturity until it is forty years old. It should by then be one hundred feet tall. The tree may go on cropping quite happily for one hundred years or more and only then will it start to decline, both in strength and fertility.

Groves of date palms always have to be planted in such a way that there are sufficient male pollen-bearers to carry out the necessary fertilizing of the female trees, for it is only the latter of course which bear fruit.

Those who belong to the Elim and Assemblies of God Churches will no doubt often refer to Exodus xv. 27 and Numbers xxxiii. 9. In Exodus it says there were twelve wells of water and in Numbers it says, "twelve fountains". In both cases, however, the seventy palm trees are mentioned and in fact the name "Elim" means "palm tree". This is the spot where one could always get fruits, for though the tree can withstand the tremendous heat of the day, as well as the cold of the desert night, it cannot exist happily without plenty of water and so the word "date palm" in the traveller's mind in the East, is very closely connected with wells and springs. The word *tamar* is common, and again and again you find names in Palestine like *Baal-tamar* and *Hazazon-tamar.* Jericho means "the city of palm trees", 2 Chronicles xxviii. 15, and Deuteronomy xxxiv. 3. The little home to which our Lord went and where he always found a welcome was at Bethany, which means "The house of the dates".

Again and again the palm was used for the decoration of the homes of the people, and Solomon thought so much of this noble tree that he used carvings of palm trees on the walls of the Holy of Holies. It may be that he did so because his father had talked about the trees being planted in the house of the Lord in Psalm xcii. 12 to which we have already referred. Even today, in the courtyards of the palaces and of the mosques one sees these graceful trees planted and here they flourish. They go on growing and growing year by year unaffected by temperatures, rains, droughts and the politics of man.

Two ancient historians refer to palms in fair detail. Herodotus states that the tree produces bread, wine and "honey". I feel that the "honey" referred to is the date palm liquor—not the honey from bees. Josephus states that there were large forests of palms by the Lake of Galilee in the Valley of the Jordan. One palm forest near Jericho was "seven miles long". It must have produced tons of dates. There certainly were palm groves round about the Mount of Olives in the time of our Lord. This was the reason it was easy for men and women to cut branches to "strew them in the way" of our Lord (Matthew xxi. 8) as he rode in triumph into Jerusalem on an ass—the colt of an ass.

The Jews had always used palms as emblems of victory, especially in their seasons of rejoicing (John xii. 13). Today some churches get dried pieces of palm leaf, twist them to form a cross, "bless them" and give them to the populace who go to the service. Presumably the idea comes from the fact that the Jews used to weave palm branches into an arch, which they fixed over the head of the bier, and this was said to speak to the mourners of victory over death and of eternal life to come.

Born again Christians will, one day, be clothed in white robes and will have branches of palms in their hands (Revelation vii. 9).

Pomegranate

The Hebrew name for pomegranate is *Rimmon* and it

is obvious that it grew abundantly in many parts of Palestine, because the name Rimmon is used frequently as part of the name of a town. Though the word "tree" is sometimes mentioned as in 1 Samuel xiv. 2, it is actually much more of a bush and is within the myrtle family. The branches are stout and thorny and the flowers bell-shaped, blood-red and the leaves, dark green. The fruits themselves are usually dull green inclining to yellow, with a blush of red spread over their surface. In Lebanon there is said to be a black variety and it is said that in Jaffa the fruits are particularly large. Normally they are about the size of an ordinary orange, but in Jaffa they can be as large as an ostrich egg!

Inside the fruits there are longitudinal compartments which contain hundreds of seeds covered with a transparent jelly-like substance. The flesh is bright pink, and the seeds are pale red. When the juiciness is squeezed out, a refreshing but somewhat bitter drink is made. This was used to produce a spiced wine, as described in Song of Solomon viii. 2. Pomegranates seem to have been grown often near great vineyards, and the juice of both may have been mixed together to form a special type of wine.

The skin of the fruits is quite tough, though thin, and sometimes the juice is used to form an indelible blue stain. The inhabitants of Palestine used it, and still do, for the tanning of morocco leather. The fruits are usually ready to use about the middle of October, but can be kept for use throughout the winter. The juice from the fruits makes a syrup called grenadine.

The pomegranate is mentioned thirty times in the Bible. First of all in Exodus, where these fruits had to appear on the hem of the High Priest's Ephod. There we read "a pomegranate, a golden bell, a pomegranate,

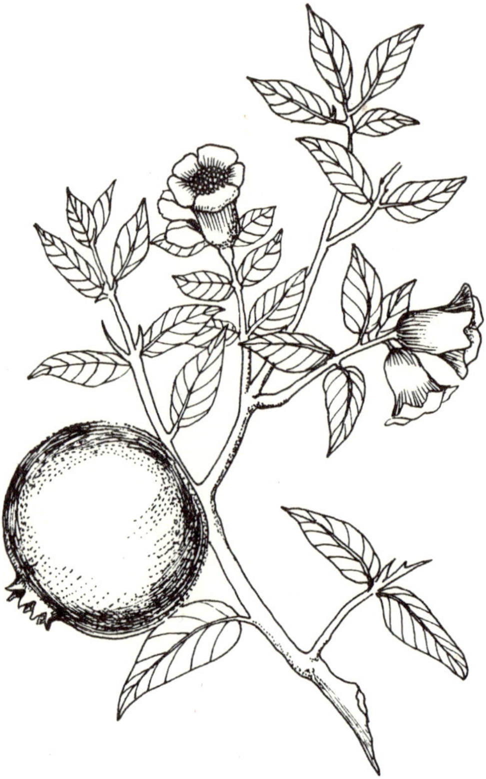

POMEGRANATE (PUNICA GRANATUM).

a golden bell" alternately. It is obvious from the context that the golden bell was the flower of the pomegranate, but instead of being red it was gold.

The spies brought back pomegranates to show them to Moses in Numbers xiii. 23, and as the fruit was greatly esteemed, we find that Moses extols the Promised Land by pointing out that pomegranates grew there in abundance.

Saul is said to have camped under a pomegranate tree in 1 Samuel xiv. 2 but it must have been an extremely large bush which gave a good deal of shade, and so he used this as a kind of headquarters. It was probably large enough to be well known, and so his men knew where to find him!

Solomon, it is said, looked at the circular calyx at the end of the fruits and used it as a model for the crown he wore. Bible students read about the pomegranates adorning the temple on the network and on the chapiters (1 Kings vii. 42 and Jeremiah lii. 22), while this fruit is mentioned again and again in the pictorial writings, in The Song of Solomon.

It is interesting to record that the English name "pomegranate" comes from the Latin names *Pomum* and *granatum* which mean "seeded apple".

Perhaps the best known of the Rimmon towns (the Hebrew name for pomegranate) was the one referred to in Syria by Naaman, who was cured of leprosy in a wonderful way, by dipping in the Jordan seven times. When speaking to Elisha afterward, he hoped God would forgive him when he bowed his head in the House of Rimmon (2 Kings v. 18) standing by his king, as he would have to do, because of his army rank. The sun god of Syria was known as Hadad Rimmon, and was the god of fertility.

Sodom Vine

Many people accept the vine "Gephen" mentioned in Deuteronomy xxxii. 32 as being an ordinary grapevine or perhaps a vine which produced, as the verse seems to suggest, extremely strong wine and, as would be expressed today, of a "rough" type. Anyway it is claimed by some that it made Lot so drunk that he was not sober for two whole days and thus allowed his daughters to commit great sin.

The writer prefers to regard this Sodom Vine as the *Citrullus colocynthis* or *Colocynth.* This is a member of the cucumber family which bears orange-like fruits and grows particularly well in the Mediterranean district. It is normally grown for its purgative properties and even today may be sold by chemists in Great Britain as Bitter Apples. This plant, which has straggling tendrils like a vine, certainly grows near the Dead Sea.

Other writers have wondered whether the Sodom Vine was really the *Calotropis gigantea* which bears short horned pods in pairs and which is known to do well in Syria and Persia. This seems unlikely, because as far as is known, a strong drink is not made from the fruits. At least one Bible student feels that the plant concerned is *Solanum sodomeum,* but this does not bear fruits like apples. It is often used for hedges in Palestine.

The Vine

We now come to the last of the fruits mentioned in the Bible and perhaps one of the most important. The vine is used as a picture of the Jewish nation. Our Lord described himself as "the True Vine" in John xv. 1. At the Lord's Supper in Matthew xxvi. 29, he made it clear that he would not drink again of the fruit of the vine until he ascended to his Father in Heaven. It was

VINE OF SODOM (SOLANUM SODOMEUM).

obvious that he knew much of the care of vines, for in Matthew xx, he described the hiring of labourers for a vineyard, while he used it in a parable in Matthew xxi.33 and Luke xx. 9.

The vine was certainly cultivated in Palestine from the very earliest days, for immediately after the flood Noah planted a vineyard (Genesis ix. 20). The Chief Butler talks about the vine to Joseph in Genesis xl. 9. The blessing of Judah in Genesis xlix. 11 includes the binding his foal "unto the vine" and we see what this means in 2 Kings xviii. 32 where it is made clear that this binding of the young animal to the vine meant—that when he got older or his descendants increased, they would arrive at the place where there would be great and profitable vineyards.

Some writers have suggested that the words, "binding his foal to the vine, and his ass's colt unto the choice vine," refers to the symbol of royalty in the house of Judah, for later on we read that "his clothes were washed in the blood of the grapes," which gives a picture of the royal vesture. It was of course, out of the house of Judah that David came, and from the line of David, we had our Lord, himself, who made clear his relationship when he said, "I am the True Vine", John xv. 1.

It is said that the Holy Land was always renowned not only for the quantity and productiveness of its vines but also for their quality. The soil and climate was right and this may be the reason why it was possible for the spies to carry out enormous bunches of grapes to prove that the land indeed was a land flowing with milk and honey.

The fact that the vine was valuable is proved by the building of a watchtower where presumably the family would reside when the grapes ripened so as to keep an eye on their vineyards. Noah may have been so impressed with the value of the vine that he took with him into the Ark a plant or two to keep until the waters subsided again. King Uzziah, who lived many years later, obviously thought that the vines were important because "he built towers and digged wells . . . and had vine-dressers in the mountains". The towers may have been for lookout purposes and would have helped in watching for unauthorized persons who might come to gather the grapes.

To the horticulturist, one of the most interesting chapters on vine growing comes in Ezekiel xvii.6 where we read, "a spreading vine of low stature, whose branches turned toward him, and the roots thereof were under him." This seems to indicate the modern method of training the young vine rods on low wires parallel to the ground. This is still done today in many of the vine-growing districts. As far as one can read in history this was the method of viticulture carried out by the Babylonians, who were very far advanced in their agriculture. Ezekiel who was an observant prophet, would, like our Lord, have seen the work going on around him and in consequence he used it in parabolic form.

The white grapes are never watered in Palestine, whereas the red grapes need constant watering if they are to come to perfection. No wonder Isaiah says, "I will water it every moment" (Isaiah xxvii. 3). When I visited the Holy Land recently I found remains of the ancient water towers in areas where the vine is now quite unknown. The stone walls or ridges which divided the vineyards in the olden days are also to be seen. The Arabs, it seems, who once occupied the Eastern slopes of Moab, used to call the ridges, "the way of the vineyards". There are similar ridges near Kadesh Barnea from which the spies brought the grapes (Numbers xiii. 23, xxxii. 8).

One of the author's favourite stories of vineyards is that of Naboth's which was coveted by Ahab for making a herb garden. One can almost hear the feet of the prophet as he strode down the path between the vines to meet the astonished Ahab and to say to him, "In the place where the dogs licked the blood of Naboth shall dogs lick thy blood, even thine" (1 Kings xxi. 19). This vineyard is said to have been on the now barren hill near the modern village of Zerin. It was on the north of Mount Gilboa that the summer palace of the Kings of Israel was built, and before the last war, it was possible to see the ancient winepresses cut into the rock. It was on this famous spot that the body of Ahab's son was cast and was undoubtedly dug into the land, in accordance with the perpetual need to nourish the soil (see 2 Kings ix. 26).

Just as it has been said that the palm can live to a ripe old age, so can the vine. The main example in this country, is the famous one at Hampton Court, which was planted there before the time of Henry VIII. It has been claimed by experts that it is by no means uncommon for a vine to be bearing for three hundred years. This gives wonderful meaning to the text in Isaiah xi. 1, "there shall come forth a rod out of the stem of Jesse, and a Branch shall grow out of his roots". When one remembers that the word "rod" really means a shining or brilliant bough, one can easily get the picture of the Lord Jesus, who came out of the stem of Jesse, like the "brilliant" rod.

Now you can cut down the cedar and it will not grow again. It has not got the power of sending up another shoot, as some trees have, and so when dealing with Assyria, the prophet made it clear in Isaiah x. 34 that the trees should be cut down with the iron axe and should not grow again. It was total extinction with them. What a difference, however, with the low-growing, beautifully fruiting vine! This could be cut down, but "a branch shall grow out of his roots".

Interestingly enough, apart from the spiritual picture of this brilliant rod representing our Lord, the wood of the vine itself is useless. It is not used today for any purpose other than for making a good fire. Ezekiel tells us this in chapter xv. 3, 4, where he says: "Shall wood be taken . . . to do any work? or will men take a pin of it to hang any vessel thereon? . . . It is cast into the fire for fuel . . . when it was whole, it was meet for no work". The cedar might be used for furniture and so could the wood of the mulberry but not of the vine. The job of the vine was to bear fruit and to bear it more abundantly.

We learn from Psalm lxxx. 13 that the boar as well as other wild beasts of the fields was a consumer of grapes. Then, of course, there is the famous story that is so often quoted in the Song of Solomon ii. 15 of the little foxes that love to rob the vines of their tender grapes, for the word "spoil" there does not mean "ruin", but is used in its old-fashioned sense of robbing, as one reads for instance in Exodus iii. 22, "spoil the Egyptians."

Of course, the vine is closely connected with wine and, though the Bible definitely condemns drunkenness, it does suggest that wine "maketh glad the heart of man" (Psalm civ. 15). In fact, right from Judges ix. 27 we discover the treading of the grapes and making merry. The drinking of wine is taken as the natural thing, just as our Lord thought it not extraordinary to provide the best wine possible for the Wedding Feast (John ii. 10) in Cana of Galilee.

The provision of a vine for every man was the promise in Micah iv. 4. In the glorious times of peace, everyone was to sit happily under his own Vine and Fig; the perfect picture of a Jewish garden with a fig for shade and a vine to produce grapes to eat and to make wine with.

To be fair one must agree that the word *yayin* in Hebrew means "What is pressed out", that is, grape juice, and there is nothing to suggest fermentation in the word. But—and this is a big but, the use of the word again and again in the Old Testament indicates that it is fermented wine about which the prophets speak. Esther i. 10, "The heart of the king was merry with wine", Isaiah v. 11 "continue until night, till wine inflame them", and again in Jeremiah xxxv. 14 "he commanded his sons not to drink wine"—no one would command a son not to drink a fruit juice! The Ancient Order of Rechabites was founded as a result of the instructions Rechab gave in this verse. This is the British Teetotal Friendly Society.

Vines and vineyards are mentioned nearly five hundred times in all. It is therefore not easy to know which references to stress and which to omit. The butler talks to Joseph about the vine in his dream (Genesis xl. 9) which has three branches, and Joseph is able to tell him that by God's grace he will go back to the Palace. The sad thing, of course, is that Scripture records, "Yet did not the chief butler remember Joseph but forgat him" (Genesis xl. 23).

One of the great punishments of the Egyptians was the "destroying of their vines by hail", so obviously, when Moses was a boy, there were vineyards in Egypt.

Vines grew in the Holy Land, in David's time at any rate, on the sides of their dwelling houses, just as they are grown in the South of England. Psalm cxxviii. 3 "Thy wife shall be as a fruitful vine by the sides of thine house". The sides of the house provide the extra warmth which enables the grapes to ripen properly. Walls give protection and support, with the result that the rods will bear heavily.

When we come to Isaiah v. we get a picture of a good fruit grower who cultivated in between his vines, who kept down the weeds and who pruned regularly, and properly. He also provided walls for shelter and protection—and hedges too. He built a watchtower so that he could see marauders and a winepress for making the wine. Despite all his care, the rods brought forth wild grapes—quite useless. There was no need to blame the soil, the situation, or the care. The trouble was, undoubtedly, "sporting" and viruses too. Sports occur in nature when you least expect them, while viruses are always with us and can turn a good apple tree into one which produces only tiny crab-apple-like fruits. Thus from huge apples one year you can get the unwanted "chats" the next.

It has been said that the Bible indicates that anyone can look after a vine, and 2 Kings xxv. 12, is quoted to prove this: "This Captain of the Guard left of the poor of the land to be vinedressers", after the siege of Jerusalem by the Chaldeans. It does not mean to say that because one is poor one is ignorant and especially in the arts of vinedressing. The farm labourer in Great Britain who in the past was poorly paid was no fool, he was an extremely skilled man—a wonderful craftsman. Vinedressing is a badly paid job and one which means being out in all weathers and on all occasions. This is seen by the reference in Isaiah xxvii. 3, "I will water it every moment . . . I will keep it night and day." Constant watering by means of the shallow ditches in between the rows of plants down which the water flowed before the stream was diverted to go to the next furrow.

Sometimes the towers built in vineyards may have been permanent, and at other times purely temporary. There seems to be concrete evidence of this latter statement in Isaiah i. 8 where we read "the daughter of Zion is left as a cottage in a vineyard and as a lodge"—the word cottage here is merely *sukkah*—a booth, a temporary structure made with brambles, reeds and the like. The word "lodge" is mentioned, which translated means "a place for passing the night", that is a temporary structure. People went and camped out in their vineyards, so to speak!

No one who contemplates vine-growing must start without knowing that there is much work to be done. There is the winter pruning and cultivation—the dunging, the spraying, the watering or overhead irrigation and the pruning and tying-in during the summer. Then there is the labour of picking, of course. The writer of the book of Proverbs scorns the lazy man: in chapter xxiv., he calls two types of men useless, ignorant men (men void of understanding) and lazy men (those that are slothful). Such men have vineyards covered with weeds of all kinds, and even the protective walls have been rendered useless by neglect.

Even the vine was to have its seventh year of rest. The vines were not to be pruned or tied that year—or even the grapes gathered; as the Bible puts it, "undressed" which really means "left alone". Fruit trees which are left alone for a year often fruit far better the following season. Many is the apple tree in a suburban garden which has been overpruned for years which the writer has coped with by advising the owner to leave it alone for a season. The result was inevitably a wonderful crop the following year on the wood which had had time and space to produce its fruit buds. God in His mercy sometimes makes his "vines" to rest.

It would seem from the information given by Josephus, the great Jewish historian, that wine was made in the Holy Land any time from the middle of September to the end of November, and that grapes for this purpose ripened at differing periods according to the aspect on which they were planted or maybe to the variety planted. There is a definite connection in Scripture between joy and gladness and a successful harvest, as for instance, in Jeremiah xxv. 30, where it indicates that the people shouted, presumably in their excitement and enthusiasm, as they trod the grapes. This shouting, as a good concordance shows us, is repeated in Jeremiah xlviii. 33, where God makes it clear that he will take away the joy and gladness from the Moabites, who would not have any wine to tread and so could not rejoice. Isaiah xxiv. 11 repeats the theme and says: "There is a crying for wine in the streets; all joy is darkened, the mirth of the land is gone".

Isaiah is full of sorrow at the loss of the vines and in consequence the lack of wine, and this is emphasized in chapter xxiv. where such sentences occur "the merry hearted do sigh"—"the noise of them that rejoice endeth"—"the joy of the harp ceaseth".

In Isaiah xxvii. 2, 3, we read, "In that day sing ye unto her: A vineyard of red wine" and, God goes on to say that He will water this vineyard every moment. Tanks of a special type for this purpose are found in the Holy Land. They are rock hewn and lined with cement and the water is conducted from them to the vine terraces by channels. These huge cisterns or dams collect the rain that falls during the winter months and at any other time of the year.

The special feast of Tabernacles was held round about the time when the grapes were gathered and pressed out in their vats, which seem to have been troughs cut out of the rocks. This was a period of giving thanks to God for the liberal harvest and was not a time for drunkenness such as took place in the heathen nations round about and, unfortunately, still takes place in some countries of the world at this season of the year.

Dedicated men like John the Baptist, Samuel and Samson, were not allowed to drink wine or vinegar or in fact the juice from any grape, neither were they allowed to eat raisins. The Nazarite vows were either for life or could be for a limited period. It is obvious that our Lord was not a Nazarite, for he said in Luke vii. 33, "John the Baptist came . . . not drinking wine and ye say, he hath a demon The son of man is come . . . drinking and ye say, Behold a gluttonous man and a winebibber".

The New Testament, in fact, is full of reference to the vine and to wine. We get the story of the wine bottles in Mark ii. 22, which are actually skins of goats, though occasionally ox skins were used if larger bottles were needed. One had to put the new wine into new skins, as then during fermentation the skin would not split but would give conveniently because it was supple.

Then there is the wonderful example of the payment of the vine dressers or the grape gatherers, who were all paid a Roman denarius, presumably, whether they started work at the beginning of the day or many hours later.

Those who worked were far better than the son

whom our Lord pointed out forcefully in Matthew xxi. 29. The message again is repentance, for the one lad was sorry that he told his father that he would not go and work in his vineyard, and he was so truly sorry that he went immediately and worked hard. The other boy said casually that he would go, and never went at all, and that is true, unfortunately, of many today.

Perhaps the most striking reference to the vine is found in John xv. where the Lord Jesus says: "I am the vine; ye are the branches." It is important to realize that when the grafting is properly done you cannot see the join and therefore one can say, as in John xvii. 23, "perfect in one". The knitting together of the graft and the stock is indeed so perfect that one can say as in John xvii. 21, "that they also may be one in us". It is almost impossible to separate the perfect union of variety and stock. A branch cannot bear unless it is fed by the roots, and the roots are there for that purpose and though John xv. 4 makes that clear, Romans xi. 18 emphasizes it, "Boast not, thou bearest not the root, but the root thee."

The Israelites in the Word of God are often described as a vine. For instance in Psalm lxxx. 8 the text says: "Thou hast brought a vine out of Egypt: thou hast cast out the heathen, and planted it. Thou preparedst room before it, and didst cause it to take deep root, and it filled the land." This is the picture of the descendants of Israel marching into the promised land under the command of Joshua, and defeating all the inhabitants, and thus settling down in cities they had not built, as was prophesied they would (Deut. iv.).

In Isaiah chapter v the nation is again described as a vineyard, which refused to produce the crop expected of it. Our Lord was continually chiding the Pharisees because they gave God lip-service, instead of life-service. As a nation, again and again the Israelites had left off worshipping the true God, and had worshipped Baal, or the golden calves which that wicked King Jeroboam had set up, 1 Kings xii. 28, 29. It took the prophet Hosea but a few words to describe the nation, for in Hosea x. 1 he says briefly but definitely "Israel is an empty vine", or to put it another way, Israel is an unfruitful or useless vine. Joel also is straightforward when he says in chapter i. verse 7 "he hath laid my vine waste, and barked my fig tree: he hath made it clean bare, and cast it away; the branches thereof are made white." He is describing the nation that was strong and without number. History records the taking of the people of Israel into captivity, and this is how God's vine (the Jewish people) was laid waste.

The Jewish nation is also pictured as a fig tree in many parts of the Word, and here we have the two "pictures" together, the ruining of the vine, and the killing of the fig tree by barking it.

Wild Vines or Wild Grapes

Reference has already been made to wild grapes as in Isaiah v. 2 and 4. The translation should really be "bad grapes". The reference is to the rebellious Israelites. There are, actually, two vines, the *Vitis vinegara* and the wild vine, *Vitis orientalis.* The latter produces tiny dry acid grapes of little value. This vine, which is useless (other than for decorative purposes), grows in Palestine today.

CHAPTER 5

THE VEGETABLES AND HERBS IN THE BIBLE

Vegetables

ONE WOULD ALMOST think, from the few vegetables mentioned in the Word of God, that the Jews in Palestine existed almost entirely on meat and fruits. There is no doubt, on the other hand, that the Egyptians grew vegetables in abundance. Some of the things the Israelites pined for when they were tramping through the wilderness to the Promised Land were the cucumbers, leeks, onions and garlic (Numbers xi. 5). It is impossible for any nomadic nation to grow vegetable crops, for they are in no area of land long enough to harvest what has been sown.

Incidentally this chapter, Numbers xi, contains several very striking pronouncements such as "Is the Lord's hand waxed short? Thou shalt see . . ." Not only did they lust for vegetables, but for meat as well, for they were tired of the manna; and God promised that they should eat meat for a whole month until it came out of their nostrils! This happened. He produced quails. Isn't "out of their nostrils" a wonderful expression? One can indeed have too much of anything.

Herbs

There are many Hebrew words translated "herbs" in the Bible: *eseb* (which means green, used for plants as a whole); *chatsir* (really grass); *botane* (herbage or shining); *deshe* (green); *yaraq* (also green); and *lachanon* is garden, or medicinal or culinary herbs (Luke xii. 42, and Romans xiv. 2).

BITTER HERBS (Taraxacum officinale).

Thus the word "herb" can be said to be used loosely in the translations. What was the prophet in 2 Kings iv. 39 looking for—some wild herb to add flavour to the stew? This same word *orah* is in Isaiah xxvi. 19, "a herb covered with dew," hence the shining herb may be a type of cabbage. I would have preferred to use the word "vegetable". But in Isaiah lxvi. 14 the "herb" really means "green tender grass." In Proverbs xv. 17 we read of *yaraq*—which I think is saladings, not herbs at all. "Better a nice salad where love is," says King Solomon.

The bitter herbs mentioned in Exodus xii. 8 were those eaten with the Lamb at the Passover. There were endive, chicory, lettuce, mint and two that we do not know at the present time. Many herbs of this sort are eaten now in Israel, and the word *Merorim* is taken to describe almost any plant that is used in this way.

Lentils

However, to get back to the vegetables and herbs, we do find extremely early on in the Bible that red pottage was made by Jacob in Genesis xxv. 34 and we know that lentils were grown, because we read of them being cultivated in 2 Samuel xxiii. 11, "And the Philistines were gathered together into a troop, where was a piece of ground full of lentils." These lentils are *adashim* or, in the Arabic, *adas.* Today, I am told, the name is *adis.* The plant which in Latin is the *Ervum lens,* or more recently the *Lens esculenta,* is a near relation of the broad bean *Vicia faba* as well as to the Tufted vetch, *Vicia cracca.* It does not grow more than eight inches high as a rule, and bears sweetpea-like flowers with leaves something like those of the Mimosa or Sensitive plant. In Palestine it is usually pulled instead of being cut, and this is done when the little pods are fully ripe. The plants are then

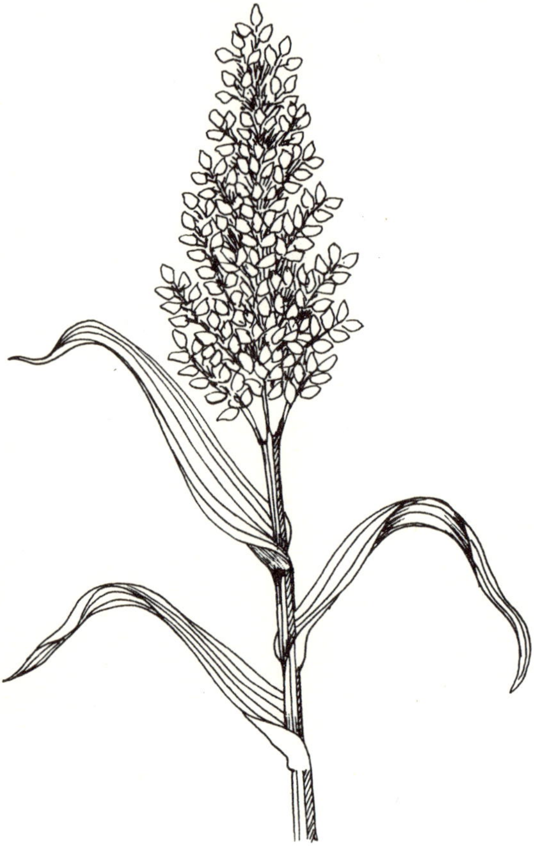

MILLET (PANICUM MILIACEUM). PANNAG.

brought to the threshing floors, where they are beaten and the little pods fall out. These then split in two and one gets the typical lentils which we know today.

The pottage made by Jacob was probably a kind of pease pudding such as is eaten in the North of England today. In the East they make a flour out of crushed lentils and this is known popularly as *Revalenta arabica.* It is supposed that the Latin name, *Lens,* is given because the two portions each look like a convex lens used for magnification. The flowers are pale blue and the plant when growing is quite attractive.

Geologists and historians have told the writer that the lentil is probably one of the oldest plants known and grown and dates back to at least the Bronze Age. Anyway, they appear a number of times in the Bible, starting with Jacob's meal for Esau, passing on to Barzillai's gift to David in 2 Samuel xvii. 28 and then on to the acreage of land growing lentils in 2 Samuel xxiii. 11 and lastly to Ezekiel iv. 9 where we find that the lentil flour, already referred to, helps to make a very substantial type of bread when mixed with the flour of barley, beans and wheat. In times of war bread is made of many different kinds of flour and comparatively recently in France, during the German occupation and afterwards, they made bread of ground maize which was always referred to scornfully as the "Biscuit Daladier".

Lentils today are still popular in Israel. It is said that red pottage is still made but today some olive oil is usually added to it, plus a herb flavouring. Whether or not the plants are still fed to cattle, I could not say, but they certainly used to be before the war and so in your lentil plant you have the dried pea for yourself and the plant for the cow.

Having thus dealt with the first vegetable mentioned

FITCHES (NIGELLA SATIVA)

in the Bible, it is necessary to turn to the herbs and vegetables found in the Old and New Testaments in alphabetical order. Some are mentioned only once, others are mentioned several times and wherever possible I have tried to give what I think is the correct Latin name and to say something about the plants as used today.

Aloe

Curiously enough it is the perfume, scent or incense that comes from the Aloe that the Bible records. Read Numbers xxiv. 6, Psalm xlv. 8, Proverbs vii. 17, Song of Solomon iv. 14 and John xix. 39.

The words used are *Ahaloth* and *Ahalim.* The perfume actually comes from the Island of Socotra which is found off the south end of the Red Sea. This is why this particular Aloe is called the *Aloe socotrina.*

Perhaps the Old Testament Aloe is really the *Aquilaria agallocha*—the Eaglewood. When the wood of this tree decays then the perfume emerges and is called Aloes Wood in Malaya and parts of China. This Sandalwood is very close-grained and fragrant. Its scent repels insects and so it is used for chests of drawers and cabinets.

When Balaam mentioned Lign aloes in Numbers xxiv. 6 he may have been referring to oaks which were known to be growing in the region.

In John xix. 39 the Greek word, Aloe, is used and this undoubtedly is *Aloe socotrina.* This was mixed with Myrrh and used for embalming. Nicodemus bought one hundred Roman pounds weight of expensive embalming perfumed powder. The fragrance of the Aloe comes from the pulp in the leaves and not from the red tubular flowers which grow in rosettes.

Anise

The plant, Anise, is recorded only in Matthew xxiii. 23, "Woe unto you, scribes and Pharisees, hypocrites! for ye pay tithe of mint and anise and cummin, and have omitted the weightier matters of the law." It was thought to be the *Pimpinella anisum,* a dainty annual plant which grows only twelve inches high and has such a very strong flavour that few people in Great Britain use it. If, however, a few leaves are cut up and put into a salad, they do add a piquant flavour.

It is obvious that the Pharisees were quite glad to pay the very small tithe of the quantity of the herb anise which was used in the kitchen, whereas they were not at all interested, our Lord said, in the judgment, mercy and faith, which our Lord described as "the weightier matters of the law". Presumably, therefore, the herb anise was used but a little, even in the Palestinian homes.

Balm

The balm of today is *Melissa officinalis,* a lovely sweet-smelling herb which grows about two and a half feet high and has rather small white flowers. But the balm mentioned in the Bible as *tsori* or *tseri* is probably the balsam which produced a medicinal gum. The Hebrew word *Bosem* is usually translated "spices" and our word, balm, perhaps stems from this. Josephus, the Hebrew historian, states the Queen of Sheba brought the balm-bearing tree as a present to Solomon, and if that is true then it means that no balm was known in Palestine until that time.

A well known hymn refers to "Balm in Gilead". Is this the *Commiphora opobalsamum*—not a native of Palestine, but certainly known and grown in Arabia? The trees were round about Jerusalem when the Turks invaded

BALM (BALANITES AEGYPTIACA)

the land, but the Crusaders reported that they had been destroyed by vandals by the time they landed to try to free the Holy City.

The balm Josephus talks about is probably quite a different tree from the one that was known to the Israelites before the reign of Solomon. It is said that the great King grew it in the tropical plain of Jericho and that it was deemed so valuable that it was one of the trophies carried into Rome by Titus the Great. Of course, it does seem that this balm was of value medicinally, for Jeremiah says in chapter viii. 22, "Is there no balm in Gilead; is there no physician there?" Now this gives some credence to the idea that it may be the Turpentine tree, *Pistacia terebinthus*, which is still to be found near Gilead today. The Arabs have always claimed that the resin from this tree has healing virtue and Josephus states that the precious balm ointment oozed out of the trunk of the tree like juice when a small cut was made with a sharp stone.

It is obvious that this balm was much prized, because when Jacob wanted to please the great Viceroy in Egypt (who was, of course, his son) he sent balm, as is seen in Genesis xliii. 11, "carry down the man a present, a little balm, and a little honey, spices, and myrrh, nuts and almonds". The Ishmaelites who were trading, and carrying Joseph with them, took balm also to sell in that country, so it probably did come from the forests of Gilead; but which particular tree the resinous extract came from, I am not prepared to say.

There is still a thorn-like bush growing near Jericho which the local folk say yields balm. Solomon certainly sought far and wide for his costly perfumes and scented plants, for in the Song of Solomon iv. 14, we read of a wonderful herb garden that Solomon had, which consisted chiefly of sweet-scented shrubs such as spikenard, calamus, camphire and cinnamon, myrrh, frankincense and aloes.

We may take it, therefore, that the balsam produces a sweet-scented mixture of an ethereal oil and a resin secreted in the hollow cavities of the plant tissue. Among the balsams used today are the Balsam of Peru obtained from the *Toluifera lereirae,* the Balsam of Tolu from *Toluifera punctata,* the Balsam of Tamacoari from *Caraipa,* Canada Balsam from *Abies balsamea* and Gurjum Balsam from *Dipterocarpus.*

Beans

Beans are mentioned twice, first in 2 Samuel xvii. 28, "Barzillai the Gileadite of Rogelim brought beds, and basons, and . . . parched corn, and beans . . . for David" and then in Ezekiel iv. 9 where the instructions are "Take . . . wheat, and barley and beans". The Hebrew word used there is *Pol* which seems to admit a translation of "pea or bean", though botanists prefer *Pol* as Broad Bean, *Faba vulgaris.* The scent of a field of Broad Beans is almost overpowering, from mid January to mid March. There is no doubt that both beans and peas were grown in Palestine and that they were used for food; firstly as vegetables, and secondly dried and used in the wintertime or ground and mixed with the flour from the wheat. Today in Jordan and Syria and Israel the principal beans are kidney beans which are either dried or eaten fresh, and a type of horse bean.

Calamus

This is another reed or cane found in Exodus xxx. 23, Song of Solomon iv. 14 and in Ezekiel xxvii. 19. The Hebrew name is *Qaneh.* This is obviously some aromatic substance extracted from a reed and is one of the ingredients of the famous anointing oil. Today there are two hundred species of Calamus, with very slender and more or less prickly stems. The climbing species grow quickly and they produce numerous suckers at their base.

There are however eighty-five species which are now called *Daemonorops* and the *Daemonorops draco* is the one source of Dragon's Blood. The author finds it difficult to be certain of the modern name of the plant to which the writer in the Bible refers.

Camel's Thorn

This is the *Aspalathus* in Ecclesiasticus xxiv. 15. "I gave a sweet smell like Cinnamon". This plant undoubtedly is the *Alhagi camelorum vas turcuron.* This is a many-stemmed thorny shrub. This I found to be well known in Palestine. It was certainly popular in the days of Pliny; he describes it and states that it is found in Cyprus—not far indeed from Israel. He describes the use of *Aspalathus* in perfume and ointments.

Camphire

This is a shrub something like a privet with the Hebrew name of *Kopher.* It is found in the Song of Solomon i. 14, "My beloved is . . . as a cluster of camphire" and the herb is mentioned again in the Song of Solomon iv. 13. Moffatt translates Song of Solomon 1. 14 as "My darling is my bunch of henna blossom" which is rather attractive. It is said to be the *Lawsonia inermis,* the Henna plant, which bears fragrant rose-coloured flowers in panicles. The plant is named in honour of Dr Isaac Lawson. It is a shrub widely cultivated in tropical countries as "the dye plant". The small oval leaves are powdered when dry and made into a paste and this is the well-known henna of India and the East which is used for dyeing. The leaves are imported into Europe and used in cosmetics. It is grown today in the West Indies and there it is known as "the Mignonette Tree".

The Arabs of the East dye their nails, palms and soles with this henna, and the shrub is said to grow largely at Engedi by the Dead Sea. The leaves are similar to those of the lilac and sometimes the blossoms are white but in all cases they are very fragrant.

In Egypt "henna" is called "kenna"—but it retains its name "henna" in Persia.

Cane Sugar

The Sugar Cane or Sweet Cane is the *Saccharus officinarum*. It is mentioned in Jeremiah vi. 20, the Hebrew name being *qaneh*, that is "the sweet cane from a far country". Some people consider that this means the Sugar Cane, and others say that the reference is to an aromatic reed from which fragrant essence is extracted. The aromatic cane or reed is usually considered to be *Andropogon aromaticus*—a grass-like plant rather than a true cane which when bruised gives out a strong scent and when eaten or sucked tastes of ginger. It is eaten by cows and goats and can taint their milk and flesh. This *Andropogon* could easily have been imported from Arabia.

The Sweet Cane could easily be "sweet smelling cane". Moffatt gives credence to this by calling it the "perfume fetched from lands afar".

In Isaiah xliii. 24 the author feels that *qaneh* here really is Sugar Cane. Should this cane be thought to be *Cymbopogon schoenanthus* then we have a densely tufted plant, found largely in North Africa and North India, which is the source of camel-grass oil. The *Andropogon nardus* is the source of the oil of citronella. In fact, the species is remarkable for the essential oils they contain which are used in perfumery.

Coriander

This is a round, aromatic seed called *gad* in Hebrew. It is mentioned in Exodus xvi. 31 and in Numbers xi. 7 where the manna which fell down from heaven was said to be like this particular seed. The coriander today is *Coriandrum sativum.* It is an annual grown for its seeds, which when ripe have a pleasant aroma. These seeds are used today for flavouring sweets, liquors, such as gin, and they can be an ingredient of curry powder or of mixed spice. In Austria, Switzerland and Germany you find Coriander seeds in bread. The seeds ripen in the month of August in this country. When unripe they have an unpleasant scent which quite disappears when they become dry. The size of the seeds gives you an idea of what the manna looked like. The Coriander grows naturally in the Jordan valley, and in many other parts of Palestine.

Cucumber

The cucumber is mentioned only twice in the Bible, in Isaiah i. 8 "as a lodge in a garden of cucumbers". Actually this is a curious translation of the word *Miqshah* because it really means "the place of the water melon", and in Numbers xi. 5 where the word is *Qishshuim* which may mean water melon, cucumber or gourd. The text however says "cucumbers and melons", and so probably it definitely means the cucumber *Cucumis sativus.* This was a very popular vegetable or salad plant in Egypt, and is still found today in Israel, planted by

CORIANDER (CORIANDRUM SATIVUM).

the acre. I am told that it is one of the main vegetables of the poor in summer time. Cucumbers grow in abundance north of Lake Genesaret, and south of Bethsaida. There is plenty of water in that area, and of course you cannot grow cucumbers without moisture.

When market gardeners grow a fair acreage of cucumbers it is quite reasonable to suppose that they may build a temporary shelter in which they may rest and sleep, looking upon this as a kind of sentry box from which they can watch the crop and prevent it from being stolen. At the end of the season the temporary shelter will of course disintegrate, the poles will fall down, and the whole thing will look ragged and sprawling. Thus you get a type of utter desolation, as in the first chapter of Isaiah.

Cucumbers were much grown in Egypt, the plants being irrigated by the Nile. The Israelites when they lived there liked them!

There is a large hairy cucumber grown in the Cairo area known as *Cucumis chate.* It looks like a melon and is commonly called the King of the Cucumbers.

Cummin

This plant occurs four times in the Word, three of which are in Isaiah xxviii and one in Matthew xxiii. 23. It is the Hebrew word *Kammon* and in the latter case the Greek word *Kuminon.* This is a herb with aromatic seeds used as flavouring. When grown in this country it is a half hardy annual sown in May, the harvesting being carried out in July and August. The species cultivated in Great Britain is the *Cuminum cyminum,* which grows six inches high, bearing white or rose flowers. It comes to us from the Mediterranean region and obviously was grown in Palestine from the earliest days. The seeds are

CUMMIN (CUMINUM CYMINUM)

used as spices in stewed meats, and occasionally it is put into bread. There is evidence that it was used as a medicine, but for what purpose the writer has no idea. It was certainly indigenous to Palestine, and was grown like any of the cereals, the seeds being beaten out with a rod.

Dill

This is mentioned in Matthew xxiii. 23 as "Anise". There seems very little doubt that it should have been translated "Dill" *(Anethum graveolens)*, the Greek word being *anethon.* This is one of the quickest plants to grow as an annual, and the liquid made from its leaves has always been regarded as useful to soothe to sleep babies and children.

In Scandinavia the leaves are used instead of mint with new potatoes, but in Great Britain the leaves are cut up finely and distributed evenly throughout a salad, and few ever discover what the delicious flavour is. Dill is also used as a flavour for pickling cucumbers. It is a little plant, grown for its aromatic seeds, which look something like those of the caraway. It grows wild in Israel, though it was cultivated in gardens. Goodspeed translates "fitches" as "dill" in Isaiah xxviii. 25 and 27 but this is most unusual.

Fitches

Fitches are a kind of ranunculus which are found growing in Palestine and produce pungent black seeds. These are used as a flavouring for cakes. In Ezekiel iv. 9 the Hebrew word *Kussemeth* obviously means prickly, but in Isaiah xxviii. 25 the word translated as "Fitches" is the Hebrew *Qetsach* which is said to mean the Black Cummin. If the plant is the *Nigella sativa* (the Nutmeg

Flower) then it grows about eighteen inches high, and bears bluish flowers (in July in Great Britain). It is known to grow happily in Israel, North Africa, and Abyssinia. The seeds are often sprinkled on cakes in Israel.

The seeds were much too small to be threshed by the usual instrument, and so it had to be "beaten out with a staff". The other species, *Nigella orientalis*, which also grows eighteen inches high, has yellow flowers spotted with red, and this was always claimed by the Jews to be the Wild Fitch, the seeds of which were used for adulterating pepper.

The best known Nigella in Great Britain is the Love-in-a-Mist which is the *Nigella damascena.*

It is obvious from the reference in Ezekiel that Fitches were sometimes used as a flavouring for bread, and especially in times of famine, when presumably they were included to make the heavy bread more palatable.

Frankincense

This is a perfume which comes from the resin of the tree known as *Boswellia.* Three types are found: *Boswellia carterii, B. papyrifera* and *B. thurifera.* The bark of the tree is peeled back and a deep cut is made with a knife; the gum oozes out and is collected in the summer.

Moffatt in his translation of Song of Solomon iv. 6 and 14 states "with all sorts of frankincense". This may mean that the gum had come from the three different species mentioned above.

Frankincense is mentioned fourteen times in the Old Testament and twice in the New. The wise men brought Frankincense in Matthew ii. 11, and in Revelation xviii. 13 it occurs in the Fall of Babylon. In the Song of Solomon iii. 6 this perfume is recorded; in Leviticus ii.1

and v. 11 we read as to how this scent was to be used.

The Hebrew name used is *lebonah* and the Greek word *libanos*—and both words mean white. When the gum is first seen coming out of the bark of the Boswellia tree it is whitish. When it is warmed and when it is actually burnt the odour is extremely pleasant.

In the case of the Roman Catholic Douai version the rendering of Song of Solomon is "All the trees of Libanus" and this is an interesting translation because it brings in the word *lebonah* with the idea that the Lebanon forest is being referred to. Could it be that the fragrant wood of junipers and pines was used when powdered to adulterate the true frankincense and thus make the total "package" less expensive? There is food for thought here.

Galbanum

It seemed to me right and proper to include Galbanum in the chapter on herbs because, if it is, as I think, *Ferula galbaniflua,* then it must come under this heading Incidentally, the plant is mentioned again in the Apocrypha in chapter xxiv. 15 of Ecclesiasticus. The experts say that this is not a Palestinian plant and, if that is so, then there is little doubt that the resinous gum was brought by the women of Israel from Egypt, and was used while the worship was centred in the Tabernacle. As in the case of other plants from which scented gum is obtained, the plan is to cut across the stem three or four inches above ground level and then to watch the milk-like sap exude. In a day or so this hardens and the gum can be picked off and used as suggested in the Bible.

There is no reason to suppose that Galbanum was used in the Temple but if it was then it was certainly imported, say, from Persia. The plants in that country grow five feet high as a rule, and the flowers produced are something like those of the Wild Parsley in this country. You get a very large number of greenish-white flowers produced in a group at the head of a short stem, and usually six or seven of these stems, only two or three inches long, at the top of a stem, say four feet six inches high.

There are a number of plants that are similarly named; there is the *Bubon galbanum,* which was found in Syria, and there is the *Opoidia galbanifera,* which grows in Persia. Knowing the Cape of Good Hope as I do, I am inclined to believe that the *Bubon galbanum* which grows there quite happily and is a native of that country could not possibly have had anything to do with the Holy Land. *Ferula galbaniflo* has many synonyms such as *Ferula persica* and *Ferula gumnosa* and these are sometimes met with in literature describing Galbanum. I am told that when Galbanum is burnt the perfume is rather strong and unpleasant, having a musky flavour. It is evident when it is mixed with other scents that its pungency is masked, and those who live in the East say that it is really used to help with the diffusion of the fragrance of other scents.

Gall

Gall in the Bible is sometimes used to describe bitterness as in Job xx. 14 or it refers to the gall bladder as in Job xvi. 13, but more often the Hebrew word *rosh* is used, meaning a poisonous herb. For instance in Psalm lxix. 21 "They gave me also gall for my meat" or in Jeremiah viii. 14 "God hath . . . given us water of gall to drink". Even Lamentations iii. 19 "Remembering . . . the wormwood and the gall".

As it is mentioned with wormwood it cannot be that herb, and as the word *rosh* is translated "hemlock" in

Hosea x. 4, "Judgment springeth up as hemlock in the furrows of the field", this makes the tracing of the particular herb concerned all the more confusing. Some experts have suggested that it may be the poppy, others have said that it is the aconite, some have even chosen the henbane, but we shall have to leave it as unknown.

The word *rosh* really means "head". This could refer to the "poppy head" from whence comes opium, but such heads as dissolved in water could become "gall to drink" Jeremiah ix. 15.

Garlic

Garlic in the Authorized Version of the Bible is spelt "garlick". It is mentioned in Numbers xi. 5, "the leeks and the onions and the garlick", as being one of the vegetables which the Children of Israel missed particularly when they were wandering to the Promised Land. It is the Hebrew word *Shum*. The garlic is of course the *Allium sativum* and it is still much used today, all over Europe, and particularly in the Latin countries where few savoury dishes are served without a flavouring of this bulbous root. It is a very near relation to the shallot, *Allium ascalonicum*, which comes from Palestine, as well as to the onions, leeks and chives.

Garlic is cultivated in Israel today and there are certain types which are found growing wild. It likes to grow in a sunny spot, and the cloves or baby bulbs are planted two inches deep and six inches apart. If they are put in during the month of February in Great Britain they are ready to lift and store in August.

Gourd

The main mention of the word gourd is found in Jonah, chapter iv. 6 where the Hebrew word *Qiqayon* is translated as "gourd". If it was a gourd, then it would be one of the types of plants which are often grown in the East climb over arbours and give shelter: a very quick-growing climber at that. In Palestine, in fact, it might grow one foot or eighteen inches in a day. Such a plant could easily be destroyed by a wireworm nibbling at its base. The gourds are usually regarded as close relations to the cucumber, their Latin name being *Cucurbita pepo.*

Many of the gourds today have been improved by selection and hybridization with the result that we have the Squashes which are very delicious to eat, there being two main groups, the winter types which are used from November to May, and the summer types from June to the end of October. The Squashes will climb up fences or up a pergola and many will bear very large fruits.

However, there is quite a different story to be told about the wild gourd of 2 Kings iv. 39 with the Hebrew word *Paqquoh* which is sometimes used to describe a wild cucumber. The story, of course, is that one of the sons of the prophets gathered the poisonous fruit of a wild vine, and it is suggested that he got hold of the *Citrullus colocynthus.* This is usually known to chemists in this country as "Bitter Apple". This plant certainly has a leaf rather like a cucumber or one of the smaller Squashes, and the powder from the pulp of this fruit is a very quick-acting purgative.

The crying out of the sons of the prophets, "There is death in the pot", was not really substantiated, because nobody actually died. The point seems to have been that it was so bitter and horrid that, as the text says, "they could not eat thereof". There is also a wild cucumber growing in Palestine which is a very strong purgative and causes those who eat it to have acute pains. How the addition of the meal by Elisha could counteract the bitterness of the vegetable, we do not know. It was regarded as a miracle. The story ends by saying "there was no harm in the pot", and this again does not indicate that the stew originally was poisonous to human beings. Perhaps, now that a large quantity of meal had been added, it ceased to have the acrid taste or the tremendous purgative effect.

Hemlock (Please also see Gall)

Hemlock is mentioned specifically in Hosea x. 4, "Judgment springeth up as hemlock" and in Amos vi. 12, "The fruit of righteousness into hemlock".

The problem lies in the two Hebrew words *laanah*, probably really wormwood, and *rosh* which the author likes to translate "a poisonous herb".

Herbs

There are many Hebrew words translated "herbs" in the Bible: *eseb* (which means green) used for plants as a whole: *chatsir* (really grass); *botane* is herbage or shining; *deshe* is green; *yaraq* is green also; and *lachanon* is garden, or medicinal or culinary herbs (Luke xii. 42, Romans xiv. 2).

Thus the word "herb" can be said to be used loosely. What was the prophet in 2 Kings iv. 39 looking for—some wild herb to add flavour to the stew? This same word *orah* is in Isaiah xxvi. 19, "a herb covered with dew", hence the shining herb: maybe a type of cabbage—I would have preferred to use the word "vegetable". But in Isaiah lxvi. 14 the "herb" here really means "green tender grass".

In Proverbs xv. 17 we read of *yaraq*—I think this is saladings, not herbs at all. "Better a nice salad where love is," says King Solomon.

Hyssop

This is mentioned eleven times in the Bible, the Hebrew word being *ezob*, and the Greek word *Hussopos.* Our Lord, for instance, in John xix. 29 was given a sponge full of vinegar, which was put "on hyssop", and in Hebrews ix. 19 when speaking of Moses it says "he took the blood of calves and of goats, with water, and scarlet wool and hyssop, and sprinkled both the book, and all the people, saying, 'This is the blood of the testament which God hath enjoined unto you'."

I find it difficult to discover what hyssop really is. There is one which is called *Zatar* by the Arabs, which grows with long slender stems. It would seem to fit into the picture described in the Bible. It has the scent of thyme about it, and it would not be difficult to make a bunch of this pungent herb in order to sprinkle the sacrificial blood. Some writers have suggested that it was probably marjoram or the Thorny Caper. As it was used in the form of a bunch or broom, it would more likely be twiggy, and this wild caper does grow in the Sinai desert somewhat in the way described. The Thorny Caper is the *Capparis spinosa,* which is a straggling deciduous shrub growing about three feet tall and covered with spines or prickles. The white flowers, which are tinged with red on the outside, always fade before noon.

Other plants that have been mentioned as the hyssop are rosemary and sorghum, while of course there is the hyssop which we know and grow today, *Hyssopus officinalis.* This bears blue flowers from June to September and has narrow aromatic leaves. If the shoots and flowers are infused in water they can be used as an expectorant. When distillation is carried out then oils are produced while are used for flavouring a number of liqueurs like Chartreuse.

One of the great texts of the Bible is that found in Psalm li. 7 "Purge me with hyssop, and I shall be clean: wash me, and I shall be whiter than snow", and for this reason hyssop is bound to have been discussed by the ancient divines. St. Augustine was very anxious that hyssop in the Bible should be regarded as a short stemmed rock plant, whose roots could penetrate deeply into the stones and between great boulders. There does seem to be something to be said for this, because in 1 Kings iv. 33 we read, "the hyssop that springeth out of the wall". If it were a tufty purgative rock plant, then this would prove an ideal little specimen to put on the end of a strong reed in order to help convey the vinegar to our Lord. One would get the idea that this little cluster of hyssop was put on the end of a strong stem like that of a bulrush and then perhaps the sponge above that. On the other hand, there are those who argue that the hyssop was used to convey the sponge, for Matthew and Mark omit any reference to this plant and record that there was just a sponge and the bulrush stem.

The author is of the opinion that the hyssop was not one particular plant at all but was probably two or three. One may have been definitely a lowly rock plant with a tufted growth and the other may have been like the rosemary with long branches well furnished with leaves that might have been used for sprinkling the blood.

Leeks

The vegetable much beloved by the Welsh, the leek, is mentioned only once—in Numbers xi. 5, as being one of the crops which the Children of Israel missed so much during their wanderings in the wilderness. It is the Hebrew word *Chatsir.* Now this word actually occurs a

LEEK (ALLIUM PORRUM).

number of times, but it is only once translated as "leek". Why it should be rendered "herbs" or "grass" in the other passages is not known, though some writers say that this *Chatsir,* sometimes rendered *Hhatzier,* is really a kind of salad plant.

"Chatsir" should really be translated "Fenugreek". This looks like a green clover. It is sold in Egypt today as *Halbeh.* The Israelis today are extremely fond of leeks and as Jewish customs and habits are handed down, it is likely that the translation in Numbers xi. 5 is correct.

There is no doubt that leeks were very popular in Egypt and there is evidence to show that they were regarded as sacred. They are certainly grown in Israel and Egypt today. The leek today is the *Allium porrum* and it is a hardy biennial cultivated for the lower part of its leaves, which are blanched. It is a cultivated species of *Allium ampeloprasum* which is found growing wild in Iran and southwards to Egypt. Actually, in its wild state, it is more like grass than a large leek, and this may be the reason why the word "Chatsir" is sometimes translated in the Bible as "grass".

The leek has been cultivated in this country since Saxon times, and in fact, in those days the leek was tremendously important. A cottage garden was called a "leac tun" just as one would call a little vegetable garden "a cabbage-patch" today. It is a plant that is supposed to have many medicinal virtues, but as far as the author knows, contains only vitamin B.

Mallows

I do not think anybody would want to eat the mallows mentioned in Job xxx. 4 unless they were very hungry indeed. If my researches are correct then this plant is the *Atriplex halimus,* which I have heard called the

Spanish Sea Purslane, or just Sea Purslane. It certainly grows in the Holy Land and especially around the Dead Sea. Visitors have seen plants growing to a height of ten feet. There is little doubt that Job could have known this plant which has a salty taste to it. It is, I believe, called by the dwellers in the Mediterranean region "The Salt Plant" today.

Even as late as the year 1600 there are records of people eating this plant and Moffatt in his translation called it "the Salt-Wort". The leaves of the plant look to me like those of the New Zealand Spinach (*Tetragonia expansa*). They were evidently dark green in colour like those of the olive, and it was the young leaves that were picked off, boiled and eaten. The plant has a tiny purple flower which opens out in the spring. You find it growing only on salty land and that is why it flourishes around the edge of the Mediterranean and near the Dead Sea. It will grow at the base of hedgerows.

Mandrake

This is included as a vegetable, because it is a member of the potato family or *Solanacae.* It bears pale bluish violet flowers as a rule, bell-shaped, and these are followed by a globose berry. The Mandrake of Genesis xxx. 14 is obviously *Mandragora officinarum,* and the fruits are always called "Devil's Apples" because of their "love potion" reputation. They are yellow and pulpy and are usually the size of a large plum. They are said to have very exhilarating aphrodisiac qualities and it is believed in the East that they stimulate fruitfulness in women. "And Reuben went in the days of wheat harvest, and found mandrakes in the field, and brought them unto his mother Leah. Then Rachel said to Leah, Give me, I pray thee, of thy son's mandrakes."

The mandrake has a root like that of the beetroot, and even today the yellow plum-like fruit is found ripe in Israel during wheat harvest, and is said to be popular on the lower ranges of Mount Hermon. The flavour is so sweet as to be sickly, and rather insipid. The Arabs claim that when eaten in large numbers the fruits produce dizziness, and may even stimulate men and women to insanity. However, there is little doubt that the desire of Rachel for Reuben's mandrakes lay along the lines of "fruitfulness". There is no proof that the Devil's Apples have any amorous properties.

In the Song of Solomon vii. 13 it says "the mandrakes give a smell". Now because *Mandragora* has no scent some people think that the plant must be *Citrus medica.*

Mint

This is the fragrant herb much used in cooking today, and found in Matthew xxiii. 23 and Luke xi. 42. In both cases, "Ye pay tithe of mint". The Greek word is *Heduosmos.* There is little doubt that this is the *Mentha spicata* originally called *M. viridis,* or perhaps the *M. longifolia,* originally called *M. sylvestris.* The latter is not so robust as the former, and in the Spring the young leaves are slightly downy. It is cultivated, because of the rapid growth of the young shoots in the Spring. *Mentha spicata* is very hardy, and is the one regarded as most suitable for the preparation of mint sauce and for other flavouring purposes.

Mustard

This is mentioned five times in the Bible; twice in Matthew, once in Mark iv. 32 and twice again in Luke. In every case it is a question of "having faith like a grain of mustard seed". As a boy I thought the seed was that

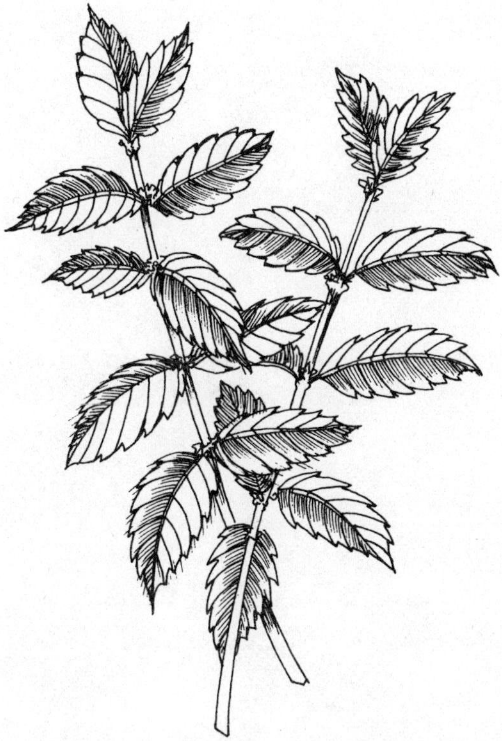

MINT (MENTHA LONGIFOLIA).

MUSTARD. (BRASSICA NIGRA).

of the mustard we grow in Great Britain as Mustard and Cress. It could be, of course, the tiny speck of a seed of the Mustard Tree I have seen again and again growing in Palestine. This is a member of the *Solanacene* family—*Nicoliana glanca,* and it does grow into quite a big tree, in which "the birds of the air could lodge".

However, Bible students as a whole feel that the seed is that of *Brassica regia,* the Black Mustard grown in our Lord's day for its content of oil. Isolated plants of this mustard can grow to a height of ten feet.

The Greek name *sinapi* has made some students feel that the plant must be *Sinapis alba* which is the White Mustard, but this never grows higher than two feet. There is also the possibility that it could be the *Salvadora persica,* the Kilnel oil plant—otherwise called the mustard tree. The author has seen the *Salvadora* growing by the Dead Sea; this plant has seeds like damson stones which could be sown by a farmer as the New Testament states.

Myrrh

There are two Hebrew words translated Myrrh, *Lot* in Genesis and *Mor* as in Psalms. Genesis xxxvii. 25 refers to "bearing . . . myrrh" and Genesis xliii. 11, "honey spices and myrrh"—is this the gum of the cistus or the ladanum?

The Wise Men from the East brought myrrh to the baby Jesus in Matthew ii and myrrh was also used at his embalming. These two facts are important as it is believed that the Holy Oil used for anointing in Exodus xxx. 23 had myrrh in it.

Now what is this myrrh? Is it the gum which exudes from the *Balsamondendion myrrha*? or is it the gum of the *Commiphora myrrha*? This latter plant is closely related to

Commiphora kataf. Both of these are shrubs whose branches when cut exude an oily sap which solidifies later.

The author feels that the Song of Solomon reference is probably to *Myrrhis odorata* which makes a good bunch of flowers with scented white blossoms. Could Solomon have grown it in his beautiful gardens? It was certainly common in Europe at the time.

Myrrh oil may have been an important cosmetic for wealthy women. Goodspeed's translation of Esther ii. 12 gives credence to this idea. Esther was making herself more beautiful and attractive to this Eastern King—rather than cleansing herself. In Revelation xviii. 11 there is weeping among the business men—pharmacists?—because of a No-Sale approach to Myrrh ointment, the Greek *Muron.* The ointment with which Mary anointed our Lord is *Muron* also—John xi. 2.

Onion

This is another of the vegetables found in Numbers xi. 5 and even today we have what is called an Egyptian onion, a form of the common onion *Allium cepa,* which produces bulbs at the top of the stem in the year of planting. These little bulbs are often found fixed with the flowers. The larger of these bulbs often produce further bulbs even higher up a new stem, so that they are given the name of Tree Onion. The smaller onions that are thus produced, are generally planted for growing into plants which produce Spring Onions for pulling green.

Onions were certainly grown in large quantities near the Nile, and still are today. They produce bulbs as large as those grown in this country, in Brittany, and France. It is believed in the East that to eat an onion is to give you some preservation against thirst. Onions in Biblical days were eaten raw, boiled, fried and roasted, and were made into soups and stews.

Pulse

The Pulse mentioned in Daniel i. 12 is *Zeroim* and in Daniel i. 16 *Zereonim.* This word is used in other parts of Scripture to describe seed and the author cannot help feeling that it is always used to describe the seed of the pea or bean family. It is obvious that soaked dried peas or soaked dried beans was quite simple food eaten by the poor and needy.

The "parched pulse" mentioned in 2 Samuel xvii. 28 is probably dried peas, but as the word "pulse" is put in the Authorized Version by the translators, one does not really know to what the verse refers.

Purslane

This is mentioned in Job xxx. 4 "Who cut up mallows . . . and juniper roots". Many feel because the Hebrew word *malluach* betokens saltiness, it may be a kind of salt-wort, a *Triplex halimus*—the Sea Purslane which grows around the Dead Sea. The leaves are silvery and pointed. The flowers are tiny and greenish in colour.

Rue

Rue is mentioned in Luke xi. 42, "Ye tithe . . . rue". The Greek word is *Peganon* and the plant seems to be the *Routa graveolens.* This is a hardy, shrubby evergreen, whose leaves have a very strong odour and a somewhat acrid taste. In this country it is used in Claret Cup. It is also a medicine used for treating cattle and poultry. It is said that four species are found in Israel today but in Great Britain I have seen only the one mentioned, which was first introduced into the gardens of this country in 1562. *Ruta graveolens* does grow in Palestine even today—but then so does *Ruta chalepensis latifolia* which is a similar plant but the leaves are divided less deeply. I am not certain to which species our Lord was referring.

Spice

Various Hebrew words are translated "spice": *basam—besem—bosem—nekoth—sammin* and in the New Testament, *G. aroma. Basam* is used only in the Song of Solomon v. i. *Besem* is used eighteen times in the books of Exodus, Kings, Chronicles, Song of Solomon and Isaiah. *Bosem* is used six times in Kings, Chronicles, Song of Solomon and Ezekiel. *Nekoth* is used in Genesis; *Sammin* three times in Exodus and *G. aroma* in Mark, Luke and John when the fragrant spice was used in our Lord's anointing.

When a mixture of spices is used as in Song of Solomon viii. 2 the Hebrew word is *regach,* while *rakal* is used to describe a spice merchant. While in Ezekiel xxiv. 10 where *regach* occurs again it is used to describe mixture of spices used for flavouring meat dishes.

Spices were, of course, used in religious services and Exodus xxx makes it clear that it was a priestly right to use them. The main perfumed spices were calamus, cinnamon, myrrh and cassia. These when powdered and mixed were sometimes stirred into olive oil. The spiced perfumes were made for worship and worship only—and by a special apothecary. The disobedience of the Children of Israel, as in 1 Samuel viii. 13 led to Samuel's warning that girls would be used to prepare the perfume unlawfully.

In 2 Chronicles xvi. 14 the King of Judah is described as

lying in a bed filled with "sweet odours and divers kinds of spices prepared by the apothecary". It is thought that his feet had gangrene which smelt unpleasantly. Spices were used to provide perfume at cremations. Ezekiel xxiv. says "Consume the flesh and spice it well".

Spikenard—or nard

This pungent scented ointment comes from the member of the Valerian family, *Nardostachys jatamensi*—the "spikenard of the ancients" as the British Royal Horticultural Society calls it.

The perfumed herb is mentioned three times in Song of Solomon i. 12 and iv. 13-14 and in Mark xiv. 3 and John xii. 3—the words "genuine nard" being used here. It was a perfume exported from India in sealed alabaster boxes; only when the guest was present in the home would the host break the seal and release the oil for the anointing.

Stacte

This is the *nataph* found in Exodus xxx. 34 which can be translated "a drop" or an "aromatic gum". In the Roman Catholic Douai version the word *cassia* is translated "stacte", Ezekiel xxvii. 19—thus it is accepted that it came from the Cassia tree. In Ecclesiasticus xxiv. 15 (The Apocrypha) the word *Nataph* is translated "sweet storax". In this case the tree is the *Liquidembras orientale.* This is the produce of the fragrant resin prepared from the bark of this tree, which is not found in Palestine.

The tree that is found in the Holy Land is the *Styrax officinalis,* a tree which produces pendulous white blooms. You have only to make a cut in the bark and the scented resin oozes out.

Wild Gourd

The Hebrew word *sadeh* really means "level spot" or "open place". So the use of the word "wild" in this connection is similar to its use in the case of "wild flowers" in Great Britain.

This wild gourd is probably the Vine of Sodom.

If the plant is *Ecbollium elatrium* then it is the squirting cucumber. But I refuse to believe that anyone would be so foolish as to gather a prickly, spiny, squirting fruit to put into a stew!

Wormwood

This is found nine times in Scripture, seven of which are in the Old Testament where the word used is *Laanah* and twice in the New Testament where the Greek word is *Apsinthos.* This latter could be translated "undrinkable". For instance, in Revelation viii. 11 we read, "the third part of the waters become wormwood", which would be better rendered, "the third part of the water became quite undrinkable".

The Wormwood is used symbolically to describe calamity and cruelty. The absinth or pastiche was obviously known in Old Testament times because Lamentations iii. 15 says "He hath made me drunken with wormwood". It is clear also that the drink is known to be unpleasant for Proverbs v. 4 says "her end is bitter as wormwood" while Deuteronomy xxix. 18 records that "the root bears gall and wormwood".

Generally speaking, however, the use of the word in Hebrew is metaphorical, though it is known that the true wormwood did grow in Palestine, being the *Artemisia absinthium,* hence the Greek word "apsinthos". In Great Britain this bears yellow flowers in August and it is known for its extremely bitter taste. Its leaves and tops have a medicinal value, and they have been used by women amongst clothes to keep away moths.

In France an intoxicating drink called absinthe is made flavoured with this plant, though *Apsinthos* in Greek means undrinkable!

CHAPTER 6

THE TREES IN THE BIBLE

THE TERM "FOREST" occurs thirty-seven times in the Word of God. Thirty-five times as *yaar*, meaning "outspread place"; once as *pardes*, meaning "park" as in Nehemiah ii. 8 and once as *choresh* in 2 Chronicles xxvii. 4 meaning "thicket".

As far as true forests are concerned we read about the forest of Hareth, south of Judah; the forest of Lebanon near Tyre; the forest of Arabia and the forest of Carmel on the south border of Asher. No one knows where the forest of Arabia mentioned in Isaiah xxi. 13 was situated.

Curiously enough the house of the forest of Lebanon referred to in 2 Chronicles ix. 16 and 1 Kings vii. 2 means the Treasure House and Armoury which Solomon built near Jerusalem. The Lebanon Forest must have been enormous. One hundred thousand foresters worked hard for fifty-five years cutting down cedars for Solomon's Temple and Palace.

It is thought that there was a date-palm forest in the Jordan Valley which stretched from Lake Gennesaret to the Dead Sea. Josephus records palm forests seven miles long in 37 A.D.

In Hosea ii. 12 we read "I will make them a forest"—this refers to forests as a whole and not to any particular species of tree. But in Deuteronomy xx. 19 we read an extremely important statement: "The tree of the field is man's life." And in Rev. xxii. 2 the Tree of Life is mentioned whose "leaves are for the healing of the nations". The indiscriminate cutting down of forests has led to deserts in many parts of the world. It leads to soil erosion, a lack of rain and so eventual starvation, and there is a very definite warning against this in Deut. xx. 19.

It seems pretty evident that trees on the whole were scarce in Palestine. Often in Scripture, one reads of "the" palm tree, "the" oak and so on, and if such trees had been abundant it would not have been easy to pinpoint a particular spot by means of one tree. One of the most interesting trees in Scripture is the palm tree under which Deborah sat. This prophetess was a remarkable woman, for she was not only a soldier, but a judge, poet and prophetess as well. She was probably an itinerant judge, as was Samuel, and presumably she lived in a tent which she pitched very often under a particular palm tree near Bethel on Mount Ephraim (Judges iv. 5).

More types of trees are mentioned in the Bible than any other plant grouping. It is said, for instance, that one can find only seven vegetables and seven flowers, but thirty-seven different types of trees are listed.

There is a lot to be said for Job as a tree lover, for in chapter xiv. 7-9 we read, "There is hope of a tree, if it be cut down, that it will sprout again, and that the tender branch thereof will not cease. Though the root thereof wax old in the earth, and the stock thereof die in the ground; Yet through the scent of water it will bud, and bring forth boughs like a plant."

Acacia

There are several species of Acacia mentioned in the Bible. One is *Shetlah.* The second is the *Acacia arabica* found in Exodus iii, the tree which yields gum arabic.

The third is the *Henora* found in Song of Solomon i. 14. Henna paste is a mixture of dried and powdered Henna leaves and the extract from the *Acacia catechu.*

The fourth is the *seneh* in Exodus iii. 2, 3 and 4. The word "bush" is possibly the *Acacia nilotica,* the Egyptian mimosa. It is thought that the name Mount Sinai comes from *Sennah.* Moffatt calls it "the thorn bush" and that is not a bad name for this *Acacia nilotica.*

Almond

This tree is mentioned in Jeremiah i. 11, 12 and in Ecclesiastes xii. 5. The Hebrew name is *shaqed* which means "hasten". This is said to refer to the fact that the flowers appear before the leaves. This is undoubtedly the *Amygdalus communis* which can grow in British gardens today. The kernels of the nuts are often eaten green, or on the unripe side before the shell hardens, or, if one prefers, the kernels are allowed to ripen fully, when they are used as a dessert or in cakes or trifles, and as a flavour, when ground, in icing.

The texts in Jeremiah i. 11, 12 make particularly interesting reading when you appreciate that almond *(shaqed)* means "hasten". (The word can also mean waken or watch.) "Jeremiah, what seest thou? I see a rod of a 'hastening' tree. Then the Lord said . . . Thou hast well seen—for I will 'hasten' my word to perform it."

In Ecclesiastes xii. 5, we read "the almond tree shall flourish . . . desire shall fail, because man goeth to his long home, and the mourners go about the streets". Now, various meanings have been given to this reference. Firstly, that the petals of the flower appear white from a distance and only pale pink when you look close. Thus you get the white hair of old age. Secondly, that the verse really means that the old men of those days were toothless (it was before the days of false teeth!) and could not, therefore, eat the nuts. Thirdly, because there were a few blossoms on otherwise bare branches, they represented the few greying hairs of an old man.

Almonds are also mentioned in Numbers xvii. 8, Exodus xxxvii. 19 and Exodus xxv. 33. In Exodus the writer refers to the "bowls being made shaped like almonds" and in Numbers to Aaron's rod that budded.

It seems from ancient Jewish records that because the rod Aaron cut and made was from an almond tree, the tree itself was blessed of God. Jehovah had caused this rod to blossom and produce nuts—while the rods of the others had remained dead. Only God can give life. This is obviously the message and of course only God can choose whom He will to lead and be a blessing to others.

Further, as all the decoration of the tabernacle was expressly ordered by God, it must be remembered that He chose that the golden candlesticks should have bowls shaped like almond nuts.

Almug

This in the Hebrew is the *almuggin.* It is mentioned three times in 1 Kings x. 11, 12: "great plenty of almug trees . . .", "the King made of the almug trees pillars" and "there came no such almug trees".

The name is also found as "algum" in 2 Chronicles ii. 8 and in 2 Chronicles ix. 10, 11. This is merely a transposition of letters, as is common in Hebrew. The tree was imported from Ophir and Lebanon and the old Hebrew writers say that it was the red sandalwood *(Pterocarpus santalinus)*—used even now in the East for making musical instruments and, in particular, lyres. It was used by Solomon for making balustrades, staircases, and terraces.

It was a tree not indigenous to Palestine. It has been suggested that the almug tree from Lebanon was too small for Solomon's liking but as he needed and evidently liked this particularly bright coloured wood, he ordered larger and better specimens from Ophir. This latter place may well have been in India.

Ash

This is mentioned only once in the Bible, in Isaiah xliv. 14: "He planteth an ash, and the rain doth nourish it . . . he maketh it a graven image, and falleth down thereto." One wonders why the translators have made this Hebrew word *oren* to be ash, for the real name should surely be pine, though the pine found in Isaiah xli and lx is *tidhar.* Perhaps the name came from the tree now known in Arabic as the Aran which is said to resemble our mountain ash. The suggestion is that it is the main tree from which idols were made. Presumably therefore the wood was hard, and when carved would not easily rot.

Bay

Most people agree that this is the *Laurus nobilis* which grows quite happily in Great Britain. It is mentioned once only in Psalm xxxvii. 35: "I have seen the wicked in great power, and spreading himself like a green bay tree." This "flourishing like the bay tree" is often quoted in common parlance today. The Bay tree is quite hardy in most places, and offers no problems at all. It grows in tubs in London, under poor conditions, and is happy in full sunshine in the country. It is an evergreen aromatic tree, which if left to itself may grow sixty feet high, the

BAY TREE (*LAURUS NOBILIS*)

leaves being narrowly oval and dark shining green, while the flowers are inconspicuous, being greenish yellow in colour.

The Hebrew word *Ezrach* which is translated "green bay tree" may also mean "native" or "indigenous", as opposed to a plant that is strange to the country or district. Some have therefore suggested that it might be the *Nerium oleander* commonly called the Rose Bay, or just Oleander. This was a shrub which was known to flourish in Palestine, by lakes and watercourses, and could grow so luxuriantly that the writer might use the simile when referring to the wicked. One finds it today growing very happily in the Mediterranean region, where it is one of the most popular free-growing shrubs, with large pink, white, rose, crimson or purple flowers.

Box

The box tree is mentioned twice, and only in Isaiah. Isaiah xli. 19 says "I will set in the desert . . . the box tree", and in Isaiah. lx. 13 reference is made to the pine tree and the box growing together. The box is known to produce very hard wood, so hard in fact that combs and spoons were made from it in the olden days. Some Bible students have suggested that as polished boxwood was often inlaid with ivory this was probably what is referred to in Ezekiel xxvii. 6. Box trees are easily grown in this country and are very hardy as a rule.

The tree mentioned in the Bible is probably the Common Box, *Buxus sempervirens longifolia.* This may grow to the height of thirty-five feet producing dark green small leaves and pale green inconspicuous flowers with yellow anthers. I am told that the box today still grows on the Galilean hills, and was undoubtedly well known when mentioned by Isaiah. It is the Hebrew

BOX TREE (BUXUS LONGIFOLIA)

word *Teashshur* which has been wrongly translated on occasions as the juniper.

Burning Bush

There are few more fascinating stories than that of Moses and the burning bush, found in Exodus iii. 2-4. Fortunately the message from the word is quite clear. The Hebrew word *seneh* does not give us much clue, because the literal translation is just "thorny bush". The same word is found in Deuteronomy xxxiii. 16 "for the good will of him that dwelt in the bush: let the blessing come upon the head of Joseph". It would seem therefore that the word used is just a wild bush of some kind.

On the other hand, when the word "bush" is mentioned in the New Testament, when referring to this incident with Moses, the Greek word used is *batos* which appears to mean "bramble bush", and four times the word is used in Mark xii. 26, Luke xx. 37, Acts vii. 30 and Acts vii. 35. One wonders whether this tree was the thorny acacia found in that part of the world. It was evidently a bush that was pretty common in the country because the Hebrew word *seneh* is said to have given the meaning to the Wilderness of Sin, as well as to the Mountain of Sinai.

Because the bush is thought to be the Thorny Acacia, some have coupled it with the Three Thorned Acacia, which is the Honey Locust, the pods of which are eaten.

Those who do not believe in God's miracles argue that the bush was probably the *Dictamus albus*—the gas plant. This is covered with tiny oil glands and so if a match is struck near by it will burst into flames. It burns for a short time and then goes out. The leaves and stems are not damaged.

It grows three feet tall, is not a bush and never burns

long enough to be the burning bush of Scripture. Others have said the bush was probably the crimson flowered mistletoe which, when the sun shines through it, may look like flaming fire. Moses was not only a Prince of Egypt but a good country shepherd so he would not mistake a parasitic plant for a real burning bush.

Cassia

One wonders about the references to cassia in Exodus xxx. 24, Ezekiel xxvii. 19 and Psalm xlv. 8. This is the word *Diddah* or *Qiddah* and seems to refer to the evergreen tree *Cinnamomum cassia* which yields the cassia bark, which was used during the war as an adulterant to cinnamon. This bark it is said was sold regularly in the market of Tyre, during the time of Solomon. It is much more pungent than cinnamon and definitely coarser. The tree is not found in Israel today, though it is still well known in South-East Asia. The inner bark of the tree is dried and it is said that it is still exported in small quantities from India.

It is interesting to note that the word *getsioth* in Psalm xlv. 8 is translated "cassia" and my concordance has it in brackets, "bark-like cinnamon". Some people have suggested that the difference is that the latter plant is really the orris root, or *Iris florentina,* though how they get at this suggestion the author cannot fathom.

Cedar

Cedars are mentioned again and again in Scripture, but only in the Old Testament. The present of cedar trees given to David by Hiram, King of Tyre, in 2 Samuel v. 11 is well known; and to Solomon in 1 Kings v. 10 and in 1 Chronicles xiv. 1. The references, however, are too many to detail, and they vary from references like "He moveth his tail like a cedar" in Job xl. 17, to "Howl, fir tree; for the cedar is fallen" in Zechariah xi. 2. The cedars were found on the western slopes of Lebanon. It is thought that the upper terraces of Lebanon were covered with forests of this noble tree and Hiram had the control of the mountains and the cedars growing there. Unfortunately there are only a few scattered groves left.

Solomon must have used literally thousands of cedar trees, first of all building the temple, and then his private palace which was so wonderful that it greatly impressed the Queen of Sheba. When you think that he sent ten thousand men a month to cut down trees, about one hundred and fifty thousand slaves to cart them, and over three thousand five hundred officers to supervise them, one can well imagine the number of trees that were felled and floated down the coast.

It took twenty years in all to build the palace and the temple, and if they were cutting down cedars all the time it is no wonder that the forests of Lebanon almost disappeared. History has it that once the trees were down herds of goats roamed the former forests, and so ate the young shoots that might have grown up to become a forest again. It is supposed that King Solomon visited the mountains on more than one occasion, for in Song of Solomon iv. 8 he says: "Come with me from Lebanon, my spouse, with me from Lebanon." It must have been a tremendous sight this one hundred thousand forest workers and the dragging of the prepared timber to the sea to float down to Joppa which was the port of Jerusalem.

Despite the fact that Solomon spent twenty-five years felling cedar trees as hard as he could go, we read that more trees were felled during the time of the return of some of the Jews from the Babylonian captivity in the time of Ezra, thanks to the interest of Cyrus, who was then Emperor of Persia. We read: "They gave money . . . meat and drink and oil, unto them of Zidon, and to them of Tyre, to bring cedar trees from Lebanon to the sea of Joppa" (Ezra iii. 7). Thus the people of those days followed almost the same procedure as Solomon, when they came to rebuild the temple of the Lord.

The Hebrew word *Erez* is translated "cedar" and the name is said to come because of the firmness of the wood. In other parts of the Bible the same word is translated "cedar tree", as in Numbers xxiv. 6 "as cedar trees beside the waters". It has been said that this word *Erez* is a generic term used to cover all the members of the pine family. Whether the cedar mentioned in Leviticus and Numbers really refers to the Cedar of Lebanon, the author very much doubts, for the true cedar did not grow in the Sinai peninsula. The reference for instance in Leviticus xiv. 4 and onwards in that chapter, and again in Numbers xix. 6, would seem to indicate that the tree here referred to was a scented type of juniper, which does grow happily in the rocky parts of Sinai. This juniper is not the modern conifer, which somewhat resembles the cedar, but probably a type of broom. (Please see further details under the heading, juniper.)

The great thing about cedar is that it is an aromatic wood with a very definite scent and taste. Insects dislike it and so do not attack it, and it is therefore free from woodworm. It has a great resistance also to fungus diseases like dry rot and wet rot, and it was used by the Egyptians for the cases of their mummies. It was also used for the construction of ships, because it would withstand the ravages of nature.

Cedars of Lebanon, of course, grow in this country to a very great height and age. The Latin name is *Cedrus libani,* and the cones do not appear before the tree is forty years of age. The trees may grow to a height of one hundred feet, and they are always objects of great beauty and dignity. References therefore like those in Psalm xcii. 12: "He shall grow like a cedar in Lebanon," refer to the strength and height of these enormous beautiful trees. The ridiculous message sent by a small thistle in 2 Chronicles xxv. 18 to an enormous cedar has far greater point when the reader realizes what majestic specimens these Lebanon trees were. When the prophet said he would take the highest branch of the cedar, in Ezekiel xvii. 3, it was a reference to the fact that it may have been growing over one hundred feet up in the air.

Chestnut

This tree occurs twice in the Old Testament, Genesis xxx. 37 and Ezekiel xxxi. 8, the Hebrew word being *Armon.* Some feel that the tree is really a plane tree and the author subscribes to this theory, because the chestnut is not indigenous to Palestine and grows there today only when imported and planted. The plane tree, on the other hand, is one of the best trees of the country. It grows huge and my Damascus friends tell me that there is a specimen there in whose hollowed-out trunk there is a shop!

Evidently it is expected to be a wonderful tree, as it is praised in Ezekiel xxxi. 8, "The cedars in the garden of God could not hide him: the fir trees were not like his branches; nor any tree in the garden of God was like unto him in his beauty." The plane tree is easy to peel and may well have been used by Jacob when laying down white wands before the ewes.

Cinnamon

The bark of the cinnamon tree is mentioned four times in all, the Hebrew word *quinnamom* being used. It is found in Exodus xxx. 23, Proverbs vii. 17, Song of Solomon iv. 14, and in Revelation xviii. 13, where the word *kinamomon* appears. It would seem that the tree itself did not grow in Syria or Palestine, but the inner rind of the bark was imported and maybe some ripe fruit also, from which the oil can be distilled. At the present time, cinnamon is used as a cure for colds, and as a flavour for food, but in the Biblical days it was much beloved as a perfume and as incense.

Even as late as the Roman times the "hippies" of those days used to scent their hair with cinnamon. The true cinnamon tree is the *Cinnamomum zeylanicum* and today the bark arrives in Great Britain from Ceylon, and is pulverized. The tree may easily grow thirty feet high and bears yellowish white flowers in terminal racemes followed by berry-like little fruits. It is a member of the laurel family, and it can certainly not be said to be ornamental.

Many fables have been built up around the cinnamon tree, because the growers in the East have no desire that their trees should be transplanted, or the seeds collected and sold. Thus in all cinnamon groves there were said to be deadly snakes or there were stories that the best cinnamon came from branches of the trees that were carried by huge birds to make their nests. One had to be therefore very valiant to be able to collect the necessary wood.

Citron

Though the citron, a member of the orange family, is not actually mentioned by name in Scripture, the Jewish rabbis state that the "boughs of goodly trees" mentioned in Leviticus xxiii. 40 refer to lemon trees rather than oranges. The actual text can be translated "fruits of goodly trees". Lemon trees on the whole have slightly larger leaves than the orange, and the flowers have a pale purple shade to them. History records that boughs of the orange and lemon trees were often used in the synagogue worship of our Lord's days because they represented God's gift of fruits to His people.

The verse Leviticus xxiii. 40 has in it two words translated "bough", the one *peri* which should be translated "fruit", and the other *anaph* as in "the boughs of thick trees", which really should be translated "branch". This word in fact occurs in Daniel iv. 12, Psalm lxxx. 10 and Ezekiel xvii. 23, whereas the word *peri* is not translated as "bough" in any other part of the Bible.

Cicero had a table, it is recorded, that was made of Citrus wood and cost him fifteen thousand pounds. It is the Sanderac tree which provides the true Thyine wood—but this is only found on the Barbary Coast or North Africa.

Cypress

This tree occurs once only in the Bible, in Isaiah xliv. 14. It was obviously a very hard-grained wood, and was used by the Egyptians, as was cedar, for making their mummy cases. The true cypress wood was very long lasting. It is thought that the word "*gopher*" mentioned in Genesis vi. 14 as the wood used by Noah for building the Ark is that of the cypress. The translators of the Authorized Version have actually used the Hebrew word *gopher* and have not attempted to translate it. Others have thought the tree might be the cedar, the cypress or the fir. Forests of cypress trees were certainly known to abound in Palestine in the very early days, and therefore it is likely that this tree was used. I am told that the Mohammedans today still like to use cypress wood for coffins because of its long-lasting properties. It is unlikely to have been a cheap wood, for it was much in demand.

Some have suggested that the tree was really the Evergreen oak, *Quercus ilex*, a wood which was much used by heathens for making idols.

Moffatt translates "Algum trees" in 2 Chronicles ii. 8 as "Cypress" and there is a lot to be said for such a rendering. Over four hundred years ago the main door of St. Peter's, Rome, was made of cypress wood and the doors are still there today. *Gopher* means a wood full of resin—so this could be the cypress. On the other hand in the United States of America I understand that gopher wood is from the tree *Ceadrastis tinctoria*—but this is not a Palestinian tree.

Ebony

This is the *Hobnim* mentioned in Ezekiel xxvii. 15. It comes from South India and Ceylon and is the very heart-wood of the date tree *(Diospyros ebenum)*. This is a large tree, the wood of which is very heavy and strong and usually the heart-wood is jet black in colour, though it is sometimes streaked with yellow and brown. It is the wood used for making the little black elephants which are often sold today, mainly in secondhand shops. Latterly, there have been cheaper imitations made of ordinary wood stained black. Very often the heart-wood of this ebony tree is only about two feet in diameter and so it is a costly article. In the Biblical days it was brought to the market at Tyre by the merchants from the Persian Gulf.

Thus *Diospyros ebenus* or *ebenestis* is really a Date palm. All translators agree, however, that the Hebrew word *Hobnim* is ebony—but it is worth noting that the Septuagint Version omits *Hobnim* altogether—it says "horns of ivory", and leaves it at that.

Fir

Normally in the Bible, the Hebrew word *berosh* is translated "fir", but there is another form of the word *beroth* in Song of Solomon i. 17. The word "fir" seems to cover the cypress, or cupressus as we so often call it now, as well as the pine. The wood from the fir trees was used as boards in shipping, as in Ezekiel xxvii. 5. It was also used for making musical instruments as in 2 Samuel vi. 5, it was employed by Solomon for making the floors of the Temple in 1 Kings vi. 15, while 2 Chronicles iii. 5 makes it clear that the ceilings were made of fir wood also. Fir was obviously used for doors and this is described in 1 Kings vi. 34. Actually this was one of the kinds of timber sent by Hiram, King of Tyre to Solomon for use in the construction of the Temple and the King's Palace.

It is difficult to be certain which is the fir referred to in scripture, but most writers agree that it is probably the *Pinus hallepensis* commonly known as the Jerusalem Pine or Aleppo Pine. Such a tree usually grows to a height of sixty feet with irregularly arranged slender branches. The cones are short stalked, and are usually produced singly, pointing backwards on a branch. This species withstands considerable periods of drought, and so grows well in the Mediterranean region, in spots that are normally too dry for other coniferous trees.

When William the Conqueror landed in England there were still fir-trees growing between Jerusalem and

Bethlehem, but during the centuries they have been badly slaughtered for firewood. Those who have said that the fir tree, described in the Bible as a "goodly tree", is the *Pinus pinaster*, have presumably built their theories on the fact that this tree grows one hundred and twenty feet high, with a long clear trunk carrying a splendid head of branches. It is found in the Mediterranean region, and was originally called *Pinus maritima*. It is the most important resin-producing pine that we have, and has been planted on sand dunes for land protection. The planting is always done when the trees are very small, and good staking has to be given until the trees are twelve years of age, for they are apt to be top heavy. It is only at the end of this period that they get a really large root system.

There is little problem about the use of the word "pine" in Nehemiah viii. 15 for there are two Hebrew words together *Ets Shemen* which being translated infer a tree of oil or very fruitful. Why these should have been translated "pine branches" is not understood, for the word *tidhar* is translated "pine" in Isaiah xli. 19, and this tree may be, as some people think, the elm which did grow on Mount Lebanon, or it could have been the plane.

We must now refer the reader to the heading "Oil tree" and the heading "Pine".

Hazel

"Jacob took . . . of the hazel", so reads Genesis xxx. 37 and the Hebrew word used there is *Luz*. The Revised Version, however, translated this "and Jacob took him rods of fresh poplar and of the almond", while the alternative to "poplar" is "storax" whose Latin name is *Liquidamber orientalis*. The name "liquidamber" refers to its fragrant resin, which is commonly called "liquid storax". The tree grows to a height of only twenty-five feet and is very slow-growing. The fragrant resin is prepared from the inner bark. If the modern translation of "storax" is right, then its use as a fertility medium is interesting, for tradition has it that Moses' staff was storax and that the scented gum used from this tree was part of the sacred perfume used by the priests.

Liquidambers certainly grow in Asia Minor and are happier in moist soils and in protected positions. There is another interesting point, and that is that the Arab name for the almond tree is *luz* and therefore it is suggested that the translation should be "almond" rather than "hazel". This almond is the *Prunus communis* which is found in the Eastern Mediterranean area, there being *amama* the bitter almond of commerce and *adulsis* the sweet almond of commerce. The normal almond of Genesis xliii. 11 and Jeremiah i. 11 is *shaqed.*

Holm Tree

The Holm tree is *tirzah* sometimes called "Holm Wood". This is the cypress of Isaiah xliv. 14. This Holm tree is the Holly, *Ilex aquifolium.* This grows not in Palestine proper—but in the Syrian mountains. It is mentioned in the Apocrypha in Susanna but I doubt if it is really a holly—it is much more likely to be the Live oak—the *Quercus ilex.*

Judas Tree

There is no Judas Tree mentioned in the Bible—but in Matthew xxvii. 5 it does mention the tree that Judas hanged himself on. This is the *Cercis siliquastrum* according to legend—and has been named such for over two hundred years. Reddish flowers like drops of blood seem to ooze out of the branches before the leaves appear. This weeping of blood each Spring is in memory of Judas. The Judas tree is a native of The Holy Land—it grows thirty foot tall—so Judas could have hung on its branches.

Juniper

The juniper is mentioned four times: first in 1 Kings 19 where Elijah lies down to die, then again in Job xxx. 4 where the roots of the juniper are mentioned as being cut up for food and being eaten by the poor. Once again in Psalm cxx. 4 the reference being "with coals of juniper", the idea being that the juniper wood was used for fuel, just as it is in Corsica and in parts of Spain by the peasants. The Hebrew word is *rothem* which should, in the author's opinion, be translated "broom" for it is similar to the modern Moorish name Retama. The broom is usually considered as *Cytisus.* The Arabic word is *Rithem* or *Retem* and this refers to a shrub which may grow to a height of ten feet thus giving ample shadow from the sun and protection from the wind.

The author considers that there is no doubt that the juniper is the *Genista raetam* which is found in Palestine and there may be called the White Broom or Juniper Bush. It is a half hardy graceful shrub, which flowers in March and April when grown in this country. It is a member of the Leguminosae family and is closely allied to the genus *Cytisus.* Propagation is best done from seed but it can be carried out by means of cuttings in sandy soil. Visitors to Palestine say that this shrub grows happily in rocky areas and that there it usually produces a straggling bush which does not cast a great deal of shade. If Elijah tried to shelter under one of these it is no wonder that he requested that he might die!

The reference in Psalm cxx. 4 obviously concerned the conversion of the juniper roots into charcoal and in the early part of this century, dwellers in Palestine claimed that the best charcoal was made from this genista root, because it threw out the most intense heat. The message in Job probably, therefore, refers to the cutting up of the mallows and cooking of these by means of the charcoal made from the juniper roots. It is unlikely that the roots of this broom or juniper would be at all edible.

Lotus Trees

The words "shady trees" are translated "Lotus Trees" in the versions by Moffatt and Goodspeed. This is the Hebrew word *Tseelim* in Job xl. 21 and 22. If this translation is correct then its Latin name is *Zizyphus lotus.* This is a bush rather than a tree which bears tiny flowers followed by round yellow fruits. It is very happy and well known in Palestine. It gives good shade it is true, but I do not myself subscribe to the "Lotus tree" translation.

Mastick

This is only mentioned in the Apocrypha Susanna 54, "who answered, under a mastick tree". The mastick is the *Pistacia lentiscus.* Gum oozes out of the trunk when it is pierced and this is called "mastick" by the trade or "mastu" by the buyers. The third grade type of gum is used for varnish. The pistacia is an evergreen which grows up to twenty feet. Fruits follow the small petaless flowers—they are red to start with, then black.

Mulberry

This is the Hebrew word *baka* and some have there-

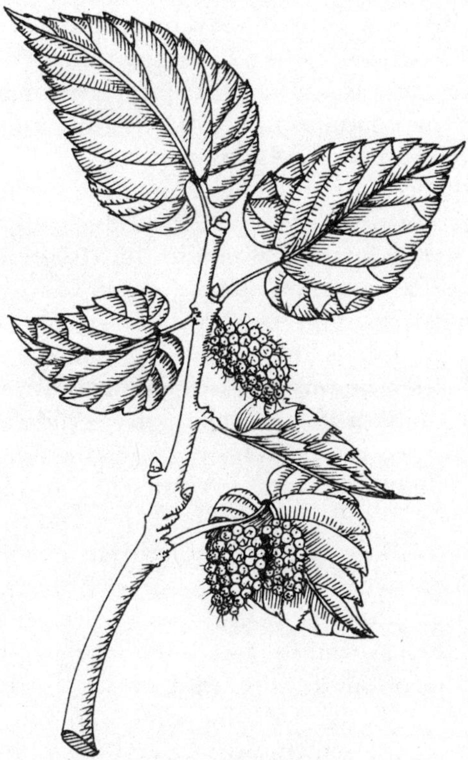

MULBERRY (MORUS NIGRA)

fore said that it ought to be translated "the baka tree". Certainly to translate the word as "mulberry" is curious. At least this has been done consistently, not only in 2 Samuel v. 23, 24, but also in 1 Chronicles xiv. 14, 15, both referring to the same incident. The Hebrew word *baka* gives a picture of weeping and therefore some have suggested that it must have been the Weeping Willow. Others have thought of a different form of weeping, that is the exudation of drops of gum and this has made them choose the balsam tree.

In this country it is the aspen poplar which makes the sort of noise suggested in the Bible reference. Several species of poplar do occur in the Holy Land, including the aspen, but then the willow is found in the lower Lebanon also. The Aspen is in fact called the Trembling Poplar and it is this trembling that David must have heard when he was told by God to "bestir himself". The Valley of Baca is "the valley of weeping" and in Psalm lxxxiv. 6 we read, "Who passing through the valley of Baca make it a well". A traveller to Zion could be refreshed in this shady valley. I am told that there is still a valley of Baca in the Sinai District. *Bacaim* of course is the plural word for *baka.*

The position is made more mysterious in a way, that "the sycamine tree" mentioned in Luke xvii. 6 is undoubtedly the mulberry, which was known well in Palestine at the time. It was grown then for its fruit, which was picked in May, the berries being exactly like those of the wild blackberries found in this country. Mulberry trees put forth their leaves late, but once they start, they leaf very quickly. The silk-worms may feed on the leaves of the White Mulberry and will produce the cocoon of silk at the end of the season.

Myrtle

The myrtle tree occurs six times, being translated from the Hebrew word *Hadas.* It is found in Nehemiah viii. 15, Isaiah xli. 19, and lv. 13 as well as in Zechariah i. 8, 10 and 11. The myrtle will be seen in bloom during the month of May in Palestine, having reddish white flowers, which are sweetly scented. The Queens of England always have a spray of myrtle in their wedding bouquets and sprigs of these myrtles are "struck" as souvenirs of the occasion. The author has seen the resulting trees growing in the garden of the Royal Lodge at Windsor.

The myrtle mentioned in Scripture is undoubtedly the *Myrtus communis* or the Common Myrtle which grows in Palestine to a height of twenty feet. It is well known about Bethlehem and Mount Hermon and on the slopes of Mount Carmel. I am told that it is still used in thc synagogues on the Feast of Tabernacles. In this country the myrtle flowers in July, producing fragrant white blossoms with numerous stamens. These are followed by purple-black berries. In eastern lands the berries are used as spices and the fragrant liquid is distilled from the flowers. The root and bark are used for tanning leather and thus they give it a fascinating smell. It was one of the trees used for constructing the booths mentioned by Nehemiah; and Milton, when describing the Bower of Paradise, also chooses myrtle. It was obviously a tree much beloved, for Isaiah when speaking of the Great Day says: "Instead of the briar shall come up the myrtle tree".

Esther's name was Hadassah, that is, Myrtle—a tribute to her beauty.

It is interesting to note that the Arabic word for Myrtle is *As* and the word for fresh is *Tur.* Therefore, "*astur*" means fresh myrtle and from "astur" you have, quite naturally, "Esther".

Nuts

The hazel rods mentioned in Genesis xxx. 37 are from the *Corylus avellana* but is it a hazel? The word *Luz* could be "almond" and there were towns in Palestine called by that name: see Genesis xxviii. 19; Joshua xvi. 2 and Judges i. 23. These towns were obviously towns with almond orchards.

The Hebrew word *egoz* is nut tree and the word *botnim* is pistacio nut—*batam* is the Arabic word. Joseph, when Governor of Egypt, received a present of pistachio nuts (*Pistacia vera*) from his father Israel. It is found in the rocky regions of Palestine bearing velvety leaves and sweet greeny yellow kernels in the small nuts. Did Pharaoh in Egypt, having seen Joseph's present, order trees to be planted in his country? The Children of Israel knew them before they left Egypt.

Oak

Oak is the Hebrew word *Elah* and is mentioned a number of times throughout the Old Testament. There is the word *Allah* in Joshua xiv. 26 translated "oak" and the word *Allon* in Gen. xxxv. 8 which is also translated "oak" or "oaks". Deborah, Rebekah's nurse, was buried beneath Bethel, under an oak, while in Zechariah xi. 2 the Authorized Version has "Howl, O ye oaks of Bashan". Lastly there is the Hebrew word *Elim* which is also translated "oaks" in Isaiah i. 29, "they shall be ashamed of the oaks". It may be that the Hebrew writers used names indiscriminately for the same tree or for what may be called different varieties of it. There is for instance the Evergreen Oak (*Quercus ilex*) sometimes called the Holly Oak and sometimes the Evergreen Holm.

Probably the most famous oak of all was the tree in which Absalom's hair was caught and this is the Hebrew word *Allah.* There seems no doubt that this was an oak, because the battle was fought in the mountainous district East of Jordan, which was celebrated for its great oaks. Another factor which seems to the author to prove that this was an oak is that Absalom's tree was part of the forest of Ephraim, and that the word *yaar* used is the same as *waar* in Arabic which means a wild region overgrown with trees, generally oaks. These facts are mentioned in detail, because some writers have asserted that the tree concerned was a terebinth, and not an oak.

Unfortunately the oak was used for hundreds of years as a place under which to put an idol, and those who visited Palestine in the 1890's still found oaks which were sacred to some spirits. It is about this that Ezekiel speaks in chapter vi. 12: "then shall ye know that I am the Lord, when their slain shall be among their idols . . . under every green tree and under every thick oak, the place where they did offer sweet savour to their idols". No wonder Isaiah speaks in Isaiah 1. 29 as follows: "For they shall be ashamed of the oaks which ye have desired and ye shall be confounded for the gardens that ye have chosen." Curiously enough, the Druids in the early days of Britain worshipped in oak groves, and held the trees sacred. The early Germanic races also believed that gods resided in oak trees.

It may be that people worshipped the oak because of its great strength, or because of its longevity. It is said that oaks live at least one thousand years, and the Parliament Oak under which Edward I held a Parliament, must be over one thousand five hundred years old. Joshua on the eve of his death set up a great stone under an old oak in Shechem. There is a celebrated old oak in Palestine, known as Abraham's oak. The spread of its branches is two hundred and sixty one feet in circumference and eighty two in diameter, and the trunk measures twenty six feet. It was there, so tradition has it, that he received the angels at his tent door, as in Genesis xviii and bid them wash their feet. It was somewhere near there that Isaac must have gone out to meditate in the field at eventide, and it was probably there that Jacob sent out Joseph to seek out his brethren in Shechem. This oak is near the Wadi Sebta and about two miles north-west of Hebron. I have seen it.

In Genesis xxxv we read of Jacob hiding his idols beneath an oak, and one wonders whether, once again, he was using the oak as a kind of receptacle for his heathen worship, instead of destroying his former gods completely. Anyway, where he buried them seems to have been considered sacred ground.

The word *Elah* is translated "Teil" in Isaiah vi. 13 and as far as I can see this is the *Pistacia terebinthus* which grows well in the Mediterranean region, and yields a resinous liquid known as Chian turpentine, which is used in medicine. If, however, we turn to Hosea iv. 13, we find this word *elah* translated "elm", presumably because oaks had been mentioned previously in the same phrase. The word *elah* occurs again as the name of the valley where David killed Goliath, and this either means the "valley of oaks" or the "valley of the teils or terebinths". Anyway, as the terebinth has the power to grow new stems after it has been felled, then this surely is the tree mentioned by Isaiah as a picture of the coming restoration of God's chosen people.

Oil Tree

This is mentioned in Isaiah xli. 19, "I will plant in the wilderness . . . the oil tree". It is the Hebrew word *shemen* which is translated "oil, ointment, or fatness". In Nehemiah viii. 15—*Shemen*, or to be correct, *ets shemen*, is translated "pine branches", while once again these same two words together occur in 1 Kings vi. 23-33 where they are translated "olive trees".

The author is of the opinion that this oil tree is probably the *eleagnus* for the Greek word *elaia* means "olive". This is a shrub or tree that resists drought better perhaps than any other and does well on a dry sunny hill or bank. This is found around Hebron and Samaria and yields a somewhat inferior oil to the olive, and bears green bitter berries. Whether, however, the shrub is the *Eleagnus angustifolia* or the *Eleagnus orientalis*, the author would not know. The latter certainly is found in the Eastern Mediterranean. Some Bible students have said that the Oil Tree is really the *Balanites aegyptiaca*, a tree which does produce oil—but the author has never heard of, or seen, this tree growing in the Valley of the Jordan. Jordanians however state that this tree provides a medicinal oil.

Pine

We must remember that the pine is akin to the fir, both being conifers. Please therefore refer to "Fir" on page 140. The translators have used the word "pine" in Isaiah xli. 19 and lx. 13, while the *shemen* of Nehemiah viii. 15 to which the author has just referred, under oil tree, is translated as "pine branches". The Latin name for the family is "pinus", coming it is said from the word "pitch", alluding to the resin which exudes from the trees. The pines are very important commercially today and many millions of pounds' worth of pine wood is imported into Great Britain each year. Pines are found in many different parts of the world, but the Jerusalem Pine, as has already been said under "Fir" is the *Pinus halepensis.*

In the early 1800s the "pine" mentioned in Isaiah was considered to be *Ulmus campestris* but this is a confused Latin name, which is now completely rejected by the experts. The name used now is *Ulmus procera* which is of course the English elm and is not indigenous to Palestine at all. The author has been unable to find an elm which is happy in the Mediterranean region, unless it be *Ulmus carpinifolia* which is the smooth-leaved elm and does grow in Western Asia.

This pine or pinetree of Isaiah lx. 13 is *tidhar* which curiously enough means "to revolve". Was this because the needle-like leaves grow in *whorls* and appear to revolve? The *etzshamen* in Isaiah xli. 19 might be the *oleaster* because it is translated "oil tree".

Though some wonder if the tree in Nehemiah viii. 15 is *Eleaeagnus angustifolia* it is translated "wild olive" in the Revised Version—was it an olive type grown for its timber and not for its fruit?

There are all kinds of stories about the pines; for instance, Josephus says that the Pines came from the Crimea and the Hiram who co-operated with Solomon fetched them by ship and did not actually grow them at Tyre.

Pistachio Nut

The tree grows in Palestine and produces clusters of lovely white flowers. It is attractive, being slender and much branched. It grows eighteen to twenty feet high as a rule and does not mind the arid areas of Palestine.

Whereas the people in the East like this nut young and fresh and will not eat it dry, the Europeans buy it readily when it is dry and like it!

Plane Tree

The Hebrew word *armon* is the Plane Tree, even though in Genesis xxx. 37 and Ezekiel xxxi. 8 the translation is "Chestnut" in the Authorized Version.

Those who agree that it is really a plane insist that it is the Oriental Plane, *Platanus orientalis.* The leaves are like those of the sycamore. It is an Eastern semi-sacred tree much valued for its shade.

There is one mention in the Apocrypha in Ecclesiasticus xxiv. 14, "I grew up as a plane tree by the water." A huge tree undoubtedly—probably seventy feet high with a trunk circumference of say forty feet.

Poplar

"Poplar" is mentioned twice: First in Genesis xxx. 37, "Jacob took him rods of green poplar", and then in Hosea iv. 13, "burn incense . . . under poplars". It is the Hebrew word *Libneh* and as the word seems to indicate "white" most people agree that this is the White poplar or the *Populus alba* sometimes called the *Abele.* This certainly is known in Central Asia, as being a tree which grows up to one hundred feet with smooth grey bark and white woolly shoots. The leaves have a white woolly underside, being dark green above. (For further information see "Hazel".)

Some have conjectured that the tree mentioned was *Populus candicans* for this is sometimes called the "Balm of Gilead", being pleasantly fragrant in the Spring. Once again the leaves are downy below. It is said that because the Poplar, *Populus alba,* threw such dark shade and gave such privacy, people used them for their worshipping groves. Isaiah, for instance, says in Chapter lxv. 3 "burning incense under the white poplars" in Moffatt's translation ("altars of brick" in Authorized Version).

It is well known that the roots of poplars spread far and wide and draw out the water and it is for this reason that Hosea xiv. 5 says "the roots struck down like a poplar" (Moffatt)—and there is no doubt he is right.

Scarlet

This is included in the Tree chapter because the insect that provides the scarlet or crimson dye lives on the oak *Quercus coccifera*—the Kermes Oak. In Hebrew is *tola-tala-shani*. The insect that provides the dye is often called the Kermes Bug, or scale. (Arabic, *kirmiz*, crimson.)

The oak itself is evergreen—it is sturdy and never grows higher than twenty feet. The acorns are one inch long and are borne in ones or twos.

The scales can cover a young branch completely—and then in turn they cover themselves with a white fluff like cotton wool so as to prevent them being eaten by birds.

Shittah Tree

Shittim wood is mentioned twenty-six times in the Old Testament mostly in Exodus and the mention of the Shittah tree itself is seen in Isaiah xli. 19. There was evidently a forest of Shittah trees at the place where Joshua sent out the spies (Joshua ii. 1). Shittim is mentioned and this is the plural of Shittah.

This is the tree from which the shittim wood was cut. It was used for the Tabernacle and was probably the only timber available in the wilderness. For this reason, it is supposed to be the acacia which is the only tree that grows to any size in that area. It is difficult to discover which species of acacia is grown. Some writers claim that the tree in Palestine has greyish black bark covered with long sharp thorns and the wood is light and durable, being capable of receiving a fine polish. This acacia grows today on the west shores of the Dead Sea.

When in Palestine I found two acacia species, *Acacia tortilis* and *Acacia senegal*—they manage to grow in otherwise barren regions. The one most commonly seen is *A. tortilis*. It grows larger than *A. senegal* and the wood, being close grained, is used for making furniture. It was used for the Ark, and the altar and the tables in the Tabernacle.

Sycamine Tree

This is the tree mentioned in Luke xvii. 6, "Ye might say unto this sycamine tree". It is almost a transliteration of the Greek word *Sukami* which should probably be the Black mulberry which used to be called *Sycamenea* in Greece and still is today in many parts. The Latin name is *Morus nigra* and it is a tree which is quite common in Palestine. This tree is grown for its fruits, while the white mulberry produces the best leaves for silkworms (please see "Mulberry" under Fruit Trees).

The Royal Horticultural Society of Great Britain says that Sycamine is *Morus nigra* or *Morus alba*, the Black Mulberry or White Mulberry (see page 146). The silk worm industry which James I wanted to introduce into Great Britain in 1605 or so never got off the ground because the King failed to realize that the White mulberry should have been grown and he insisted that the Black mulberry be planted.

This White mulberry will grow to a height of thirty five feet and give thick shade. It was introduced into Palestine only one hundred years ago—so the tree in Our Lord's time must have been *Morus nigra.* On the Island of Crete I found that this mulberry was called a Sycaminos!

Sycamore

The Hebrew word *shiqmah* and Gr. *sukomoraia* are translated "Sycamore". So we read of the Sycamore in the Old Testament in 1 Kings x. 27 and Amos vii. 14, but the story of Zacchaeus climbing up the sycamore in Luke xix. 4 is the best known mention of this tree.

A very hard resin exudes from its trunk and branches, which is used in the manufacture of varnishes. This is known as assandarac. The Romans called the wood citron wood and liked it immensely. It is not normally found in Palestine, but grows on the Atlas mountains.

This is occasionally thought to be the mulberry tree also. Most Bible students, however, insist that this is no mulberry tree, but the *Ficus sycomorus* or Sycamore Fig, which is abundant in Palestine and Egypt.

It has a leaf like a mulberry but a fruit like a small fig, and is an evergreen. It can grow to a large size and the timber from this tree was commonly used by the Egyptians of the Pharaohs' time, for furniture, boxes and doors. It is commonly found in the plains of Jericho. It belongs to the family *Moracae* and therefore is very closely related to the mulberry. It is an easy tree to climb and is often planted by the roadside, so that it would be an ideal tree from which Zacchaeus could look down directly upon our Lord. The tree is said to bear seven crops of figs a year in Palestine and is easily propagated by cuttings.

The wood of the sycamore tree, however, is soft and of little value, and that is probably why Isaiah says in chapter ix. 10, "the sycamores are cut down, but we will change them into cedars." That is why, also, in 1 Kings x. 27 Solomon is said to have made cedars of Jerusalem to be as the sycamore trees for abundance. The great prophet Amos was a cultivator of the sycamore tree, presumably for its fruits—Amos vii. 14, "I was no prophet, neither was I a prophet's son; but I was an herdman, and a gatherer of sycomore fruit." People were appointed to take care of these trees, as one can read in 1 Chronicles xxvii. 28. In Psalm lxxviii. 47 the sycamore trees are destroyed by frost. The fruits of this sycamore fig are still eaten today.

Amos was a gardener as well as a herdsman and he knew the importance of pricking each fruit with a sharp skewer or knife at the right stage in its development, in order to help it to ripen properly. The word "gatherer" used in Amos vii. 14 really means "one who cuts or scrapes"—and he cut into the little figs. The Roman gardeners knew about this gardening operation and Pliny actually refers to it in his writings.

Tamarisk

In the Authorized Version Tamarisk is not mentioned—but Goodspeed and Moffatt insist that the word "shrubs" in Genesis xxi. 15 and "tree" in Genesis xxii. 6 and xxxi. 13 should be translated "tamarisk". This is a tree that would grow in a droughty district and so would be quite happy in the Desert of Shur where Hagar found herself, and it would give good shade to the baby Ishmael, when she laid him down.

The Teil and Terebinth

The Hebrew word *elah* has been translated "oak, plane, teil and elm". It is probably the *Pistacia terebinthus* which is very common in the south and east of Palestine, in places where it is too dry for the oak, which it closely resembles. The terebinth is said to bear small clustering blossoms and red berries and many is the Bedouin sheik who has been buried under such a tree. Sometimes the fruit is the same size as the pea and in at least one variety it is four times as large. Oil is produced from these berries by crushing them like olives. In the case of the true *Pistacia terebinthus*, however, the fruits borne are more like nuts, three-quarters of an inch in length, which are very pleasant to eat.

In Joshua xxiv. 26 the tree is translated "terebinth". In Isaiah vi. 13 it is called "oak" and in Hosea iv. 13 it is translated "elm".

The true translation of *Elah* is "a strong hardy tree". It is thought that the Children of Israel liked to sacrifice under the terebinth tree because it threw a black thick shadow—you need darkness for sinning as a rule!

David slew the giant Goliath in the Terebinth valley. Did the Terebinths produce such shade that Goliath did not see the slung stone easily? This *Pistacia terebinthus palaestina* is a common tree in Palestine growing often thirty-five feet high.

This tree provides the Cyprus turpentine which oozes out of the branch when it is incised. The specimen is called the Turpentine tree in Ecclesiasticus xxiv. 16 (The Apocrypha): "As the turpentine tree". It looks like an oak in the winter and an ash in the summer!

God evidently is pleased with Terebinth planting. "They might be called the Terebinths of righteousness —the planting of the Lord, that he might be glorified".

Thyine Tree

This is found in Revelation xviii. 12 where the Greek word *Thuinos* appears. It is said to be a small tree with the Latin name of *Callitris quadrivalvis* which has recently been renamed *Tetraclinia articulata.* This is a very tender evergreen tree, which grows about fifty feet high. The wood is hard, fragrant and prettily marked and therefore is much esteemed by cabinet makers.

Walnut

This is a tree that flourishes even today in parts of Palestine—or Israel, as it should now be called. It obviously grew abundantly in Solomon's time, as Song of Solomon vi. 11. The Hebrew word is *Egoz* and it presumably is the *Juglans regia.* The word for "nuts" used in other parts of Scripture is *botnim.* Solomon evidently had a big garden of walnuts. These, of course, may have been comparatively small trees as are grown in some parts today, or the very large specimens known in many parts of Great Britain.

"I went down into the 'market' garden of walnuts" —or maybe a "wood of walnuts". Song of Solomon vi. 11. After all, the gardens of those days were beloved for their shade. I, therefore, imagine a lovely shady area where paths wandered among those beautiful trees and where, in the autumn, the nuts might be picked.

Willow

The Hebrew word for "willow" is *ereb* and it is found five times in the Bible in Leviticus xxiii. 40, Job xl. 22, Psalm cxxxii. 2 as well as in Isaiah xv. 7 and xliv. 4. Perhaps the best-known reference is that of the 137th Psalm: "We hanged our harps upon the willows." A willow tree is mentioned in Ezekiel xvii. 5 as *tsapht-*

saphah. The second willow was used for constructing booths in the Feast of Tabernacles, and was obviously a tree that grew by the waterside.

Some consider that it may be the oleander or Rose Bay (*Nerium oleandea*) which is found flourishing very happily by the watercourses of Palestine. The weeping willow is the *Salix babylonica* which came to England from the region of the River Euphrates, while the other willow is probably *Salix medemii* which is found in Armenia and Persia today. It is one of the earliest willows to flower.

The weeping willow mentioned above is as I have said the *Salix babylonica* but in my researches I cannot find that it was ever grown in Palestine—but as there were so many other kinds of willows growing in the Holy Land we must only conjecture as to the name of the actual tree mentioned.

Willows are always associated with sorrow, and there was a popular song of the Victorian days with the words "I'll hang my harp on the weeping willow tree, and never, never think of thee." This was obviously a reference to Psalm cxxxvii. On the other hand the willow can be used for comfort, for God says when speaking of his church "and they shall spring up as among the grass, as willows by the water courses" (Isaiah xliv. 4), thus indicating that God will ever provide the "water of life" to his people.

Samson was bound with green withes and these undoubtedly came from willows, Judges xvi. 7; and in Amos vi. 14 we read of "the river of the wilderness" —which could be "the willow river" or, as Moffatt puts it, "the torrent of willows". Moffatt also uses "Poplar" when translating *Ereb* in Psalm cxxxvii. 2. This is the *Populus euphratica.*

CHAPTER 7

THE FARM PLANTS IN THE BIBLE

TODAY, AGRICULTURE is poles apart from Horticulture. There are agricultural advisers and horticultural advisers, there is the Royal Agricultural Society and the Royal Horticultural Society; the Horticultural Education Association and the Agricultural Education Association—and so on. In the Bible days, however, any work done on the land was just farming and so no one attempted in those days to differentiate between the purely agricultural and horticultural.

This chapter, however, deals with the plants one would normally find on an agricultural holding rather than a horticultural one. The general cultivations necessary on a farm will be found in Chapter 8. All we have to do, therefore, in this chapter is to list and write about the plants mentioned definitely in the Word of God—and give what details we can about them.

Barley

Barley is mentioned again and again in the Bible. We first find it in Exodus ix. 31: "the barley was smitten; for the barley was in the ear". Then we see that a homer of barley is valued at fifty shekels of silver in Leviticus xxvii. 16. It was a loaf of barley bread which, in the man's dream, tumbled into the host of Midian, knocking down a tent, in Judges vii. 13. Naomi returned to Bethlehem bringing her daughter-in-law Ruth at the beginning of the barley harvest (Ruth 1), whilst of course, the gleaning Ruth did was barley gleaning and the "corn" Boaz gave her as a present in the third chapter of Ruth was barley.

When Absalom planned to annoy Joab it was his barley field that he burned with the aid of his servants. 2 Samuel xiv. 30. The well-known text, "Cast thy bread upon the waters: for thou shalt find it after many days",[1] refers to barley bread and not to wheat bread (Ecclesiastes xi. 1), while the boy who gladly gave up his food to the Lord Jesus in John vi. 9 had five barley loaves and two fishes.

Barley loaves were the food of the poor and this is made quite clear in the Bible because for the highest offering it must be a loaf of fine wheat flour, as in Leviticus ii. 1 whereas we read in Numbers v. 15 it was sufficient to bring an offering of barley for an unfaithful wife. The price of barley flour is said to be half that of wheat flour, as Elisha made it clear in 2 Kings vii. 1 "a measure of fine flour . . . for a shekel, and two measures of barley for a shekel".

Solomon must have grown or bought a tremendous tonnage of barley each year for it seems, if we read 1 Kings iv. 28 aright, that he fed his horses and dromedaries with barley, and as he had forty thousand horses, at least, and an unknown number of dromedaries, they must have eaten a vast quantity of barley or barley meal.

The Latin name for barley is *hordeum* and the Hebrew name *seorah*; this is interesting because *seorah* means long-haired, which, of course, describes the awns or long bristles that you find on the ears of barley. The Greek word is *krithe* which means piercing or pointed,

[1] Moffatt renders this passage "Trust your goods far and wide at sea till you get good returns after a while."

and that again refers to the long bristles. It is supposed that three types of barley were grown and that the thirty-two references to barley, barley meal or barley loaves may refer to (a) the spring barley *Hordeum vulgare,* (b) the winter barley *Hordeum hexastichum* and (c) the common barley *Hordeum distinchum.*

As far as I can discover the sowing of the winter barley in Palestine was probably done between the end of October and the middle of November and as a result the harvesting could take place in time for the Passover. The spring barley was then sown when all sign of winter was over, and was harvested in the summer. It depends, to a great extent, where the field is in Palestine, as to the exact time of the harvest, but the Israelis today can harvest the winter-sown barley any time between March and May. Fortunately, barley is a cereal which will grow under very difficult conditions, and even grows quite well in times of drought when wheat would suffer badly. Further, it is said that in the northern part of Palestine the summers were not long enough to ripen wheat and so barley had to be grown.

Barley was an important crop in Egypt and I learn from official sources that one can usually reckon to harvest barley today in most parts of Egypt thirty or more days before wheat. As this was probably the case in the times of Pharaoh, it accounts for the results of the plague in Exodus ix. when in verse 31 we read "The barley was smitten" because, of course, it was ready for harvesting, but the wheat was not touched because the grains were not in the ear; it was still only half-grown.

One would have found barley grown extensively in the valley of the Jordan, in and around Lebanon and in the land of Moab, while today it is much grown in Syria.

I am told that the Arabs today refer to the Jews as

"cakes of barley". This is of course a scornful term and indicates that the followers of Mahomet regard themselves as "the wheat". Maybe the Midianites used the same term for the Israelites in the days of Judges and that accounts for the destruction of the tent by the "barley cake" that we have already mentioned (Judges vii. 13).

Though today barley is much used for making intoxicating drinks it was never used for that purpose in Biblical days. Barley was there as a food and if an alcoholic drink was needed it was made from the vine. Though in Europe we eat wheat bread, almost entirely, the food of the poorer people in Palestine was, and still is, barley bread.

Bulrush

The bulrush is included in this chapter because, though it is not cultivated in the normal sense of the word, it was used for making baskets and even paper or papyrus on which the Egyptians wrote much. The most important reference in the Bible is undoubtedly that concerning baby Moses, because it is said that his mother made a little boat with the bulrushes and placed him in it. It was probably a little boat because, in Isaiah xviii. 2, it refers to "vessels of bulrushes on the waters". The weaving together of the papyrus bulrushes to form the little boat must have been carefully done, and in order to make it absolutely waterproof it was daubed inside with thick mud called by some translators "pitch", which was undoubtedly allowed to harden, and outside with pitch or tar.

To make the paper from this particular bulrush the Egyptians extracted the pith very carefully, laid it flat on a stone and then glued the outside strip of one flattened

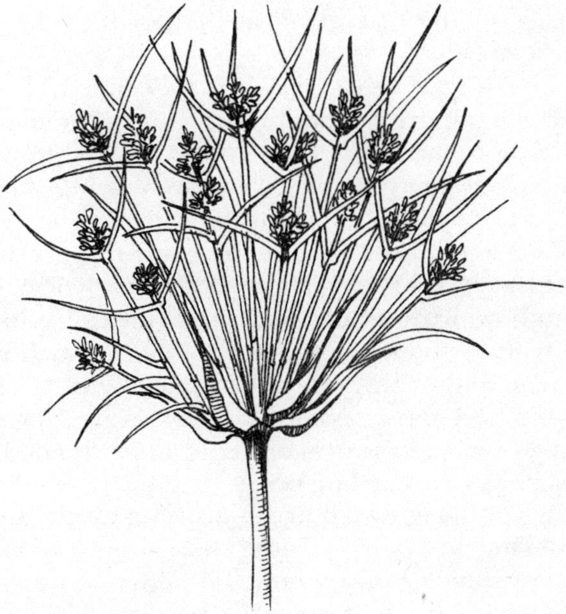

BULRUSH (CYPERUS PAPYRUS).

piece of pith to the next. Then when the paper was made to the right size, very heavy weights were put on top of it to press it absolutely flat and level and later it was dried. It was at that point, of course, that it could be used for letters and messages. There were many areas of the Nile that are supposed to have been completely covered with these bulrushes in the time of the Pharaohs, just as I was told that before the war it was a perfect menace in the marshland at the north of the Plain of Gennesaret.

This tall-growing bulrush looks very much like a mop-head when it grows, though perhaps far more graceful. It has a tremendous root system which prevents the plant from being washed away by moving water. One can imagine these mop-heads bowing down in the wind looking like a crowd of people, and this apparently is what Isaiah is thinking about in chapter lviii. 5. He talks about bowing down the head like a bulrush as indicating formality without faith and penitence. Isaiah referred often to bulrushes; when he talks about the parched ground becoming a pool in chapter xxxv. 7, he says that the place where the dragons used to lie should be water with its reeds and bulrushes.

If one can imagine a bulrush growing twenty feet high, producing a very thick horizontal stem, then it is not difficult to see how the paper was made from quite a wide piece of pith. If one notes that it was possible to eat the pith of the bulrush, either cooked or raw, then this may give the answer to the query that has sometimes been raised in connexion with Ezekiel iii. 1, where God says "eat this roll" and in verse 2 where we read "he caused me to eat that roll" and once again in verse 3 "Then did I eat it; and it was in my mouth as honey for sweetness". Ezekiel was referring of course to the scroll or roll of a book mentioned in Ezekiel ii. 9. Though the passage can be spiritualized it is interesting to know that the scroll of those days would be papyrus and the papyrus could be, and was, in fact, eaten.

One reads of the strong thick roots of the papyrus being used as handles; one finds the thinner stems of the bulrushes interlaced or woven in order to make mats and it is reported that the bulrushes were often made into shoes.

Abyssinians say even today the lower part of the bulrushes are chewed by the poorer children and that they have a taste of licorice to them, while certainly small boats are made by young people out of the stems of papyrus, just as it is said that paint brushes were produced in the time of the Egyptians by using a small bulrush stalk and just fraying out one end very carefully. The bulrush is, of course, *Cyperus papyrus* and this is undoubtedly the plant to which Isaiah refers in chapter xix. 7 "The paper reeds by the brooks".

Cotton

Cotton is included even though it is only mentioned once and that in the Book of Esther (where God is never once mentioned). I have had to turn to the Revised Version to find the use of the actual word, for in the Authorised Version we read in verse 6 of chapter i of the Book of Esther, "white, green and blue hangings", but in the Revised Version, "white cotton curtains and blue hangings". It is recorded that cotton was grown in Judaea round about 480 and 490 B.C. and comparatively recent modern discoveries have shown that some Egyptian mummies are wrapped in a cotton material.[1]

[1] It was usually the children who were embalmed in cotton bandages.

It is said that the plant grown in Palestine was *Gossypium herbaceum* and that the cotton it produced was somewhat yellower than that grown in the Southern States of America.

The cotton bushes grew about five feet high and normally bore pinkish flowers and I have reason to believe there was a yellow-flowered variety, as there is in China today. The pods that follow the flowers burst when they ripen and it is out of these pods that the fluffy cotton is seen. When the fluffy masses, or bolls, are gathered they are then processed and woven.

It is interesting to note that the type of cotton used today in this country may receive its name from the place from which it originally came. The word "calico", for instance, comes from the town of Calicut in India, and the word "muslin" from the town of Mosul in Turkey. The Persians certainly used much cotton to make materials and the Jews would have learned cotton weaving when they were in captivity under the famous Ahasuerus of the Book of Esther, for he reigned, you remember, from India, even to Abyssinia, over one hundred and twenty-seven provinces.

It is suggested that Isaiah makes reference to cotton in chapter xix. 9, where he says "they that weave networks". This word "networks" is translated "white-works" by some and "white cotton" by others.

Flax

Most people in Great Britain will know of flax as a plant, for it is very pretty and dainty. It can grow to a height of three feet though in this country it is usually two feet high and the flowers are a lovely blue and they have a violety-blue marking at their base. In some parts of Great Britain one can see whole fields of flax in flower,

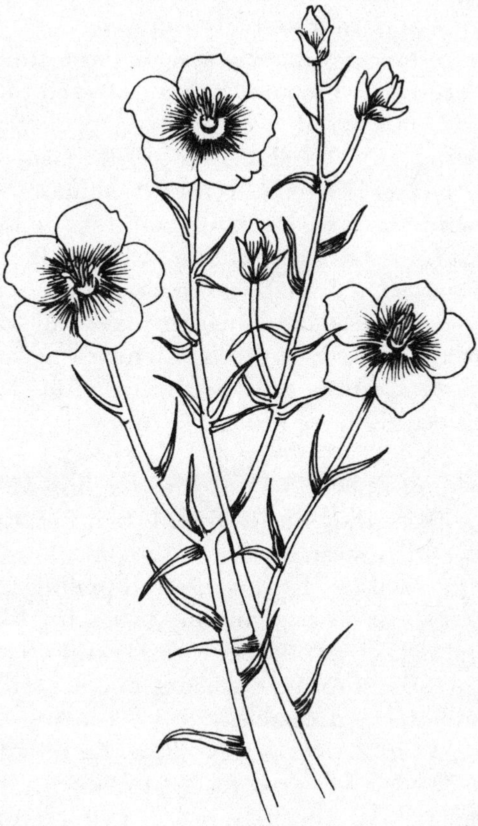

FLAX (LINUM USITATISSIMUM).

and they look very pretty. It is said that the spies found flax growing in the land of Canaan long before the Children of Israel actually entered.

The common practice was to grow the plants until they were fully ripe and then they were pulled up whole and laid out to dry. This was probably what was happening on the flat roof of Rahab's house when Joshua's spies arrived. I found difficulty as a boy in understanding how the spies could be hidden on what I thought must be a very thin matting of flax on the roof, but of course, if Rahab was drying a fair acreage of flax from her plot of land, which may well have lain just outside the city, then the flax which was laid out to dry on her roof could easily be three or four feet thick and the spies could have hidden in this without any difficulty at all.

The next operation was the steeping of the stems in water so that they decayed and then Rahab would have had to split the stalks and use some kind of comb in order to separate the threads and peel them away. This operation was known in the 1700s in this country as "retting". Next, of course, Rahab would have woven the fine threads into linen using the distaff method of spinning in all probability.

We read in Proverbs xxxi. how the Jewish women did the work—for in verse 13 she works the flax "willingly with her hands" and then in verse 19 she puts her hands on the spindle and the distaff. So we go on to verse 24 where we read, "She maketh fine linen, and selleth it".

Of course we know that the linen was used in a dozen and one ways for we find it making the ceremonial robes worn by the priests in the Temple, while again we discover in Isaiah xlii. 3 that it was even necessary to use linen for the wicks in the lamps or candles. To do

Joseph honour Pharaoh arrayed him in fine linen (Genesis xli. 42). The Tabernacle had ten curtains of fine twined linen (Exodus xxvi. 1). The priest was to wear linen breeches, as in Leviticus vi. 10 and it is fascinating to note the definite instruction in Leviticus xix. 19 that garments were not to be made of mixed wool and linen. In fact this instruction is repeated with emphasis in Deuteronomy xxii. 11, and many people today, say that a mixture of cotton and wool is more irritating to the skin than either of these materials alone.

Fine linen was evidently an extremely important material because Mordecai went out from the presence of the King in royal apparel and this consisted of a great crown of gold and a garment of fine linen Esther viii. 15. Even sails were made of linen, we note, in Ezekiel xxvii. 7, and not only were these linen sails fine, but they were evidently beautifully embroidered. When Joseph took the body of our Lord, he wrapped it in a clean linen cloth (Matthew xxvii. 59, and again Mark xv. 46, Luke xxiii. 53, and John xix. 40).

It is fine linen that the "saved" will wear in heaven, Revelations xix. 8 "for the fine linen is the righteousness of saints".

There seem to have been three types of linen used in the times of the Bible. One, presumably, of a coarse texture, as in Ezekiel ix. 2, and Daniel x. 5, one of a better texture as in Exodus xxvi. 1. It is obvious that the failure of the flax was important to the Children of Israel for God mentions it as one of his punishments in Hosea ii. 9.

I would like to be able to read the Talmud, much beloved by the orthodox Jews, for I am told that it contains quite detailed instructions about the way flax should be harvested, bleached and prepared, presumably with special reference to the making of the robes to be worn by officiating Rabbis.

The Latin name of flax is *Linum usitatissimum* and it is undoubtedly the oldest of the textile fibres.

Grass

I think the word "grass" occurs fifty-eight times in the Old and New Testaments. Sometimes in a practical farming way, as in Jeremiah 1. 11 "fat as a heifer at grass" and other times pictorially as in Isaiah xl. 8. "The grass withereth and the flower fadeth—but the word of the Lord shall stand for ever," or as in Psalm ciii. 15 "As for man his days are as grass."

I like the straightforward statements like "grass for grazing" in Job vi. 5 or as in Psalm xxxvii. 2 "they shall soon be cut down and wither as the hay," for the green herb is "hay".

The Hebrew words translated as "grass" are *deshe* (which really means tender) *yereq* (whose real meaning is green), *Eseb* or *Asab* is "herb". When turning to the Greek we find *chortos,* meaning "fodder".

We find Our Lord in Matthew using "grass" to describe plants as a whole. "If God so clothes the grass in the field . . . shall he not much more clothe you". I feel he was speaking of flowering plants—especially lilies.

The Apostles agree with the Prophets and Psalmists for in Psalm ciii. 15 we read, "as for man his days are as grass," and in 1 Peter i. 24 we find "All flesh is as grass and the glory of man as the flower of grass." James i. 10 continues the theme when he says, "As the flower of the grass—he shall pass away".

It is thought that the Children of Israel used Sorgham grass, *Sorgham vulgare,* on the first morning of the Passover, as to have obtained the large quantities of real

Hyssop would have been almost impossible. The Sorghum grass was used in those days to make brushes so it would be quite suitable for the purpose here.

Manna

This is included as it was a miraculous food provided by God for the Children of Israel in the wilderness where they were unable to grow their own wheat and barley. It makes it clear in Numbers xi. 6 that they had "Nothing at all, besides this manna" which looked like "Coriander seed", Numbers xi. 7. In Deuteronomy viii. 3 Moses made it clear that "they were fed with manna which thou knewest not". He also said "that He might make thee know that man doth not live by bread alone—but by every word that proceedeth out of the mouth of the Lord." The Lord Jesus quoted this to Satan at the time of his forty days and forty nights in the wilderness—see Luke iv. 4.

The Psalmist in Psalm lxxviii. 24 says "it rained down manna upon them to eat", while John's gospel says in Chapter vi. 31 "Our fathers did eat manna." In Hebrew the word is *Man* and in Greek—*manna.* In Hebrews we read in Chapter ix. 4 of the putting of manna into a golden pot thus referring to Exodus xvi. 33 in which Aaron, the High Priest, is depicted putting an omer of manna into a pot (presumably golden) and placing it "before the Lord in the Tabernacle." Thus the generations to come would be able to see the bread that God provided free in the wilderness.

This manna tasted evidently like "wafers made with honey"—see Exodus xvi. 31.

Revelation ii. 17 says "to him that overcometh will I give to eat of the hidden manna"—a spiritual experience evidently.

Pannag

There is little doubt at all that the Pannag mentioned in Ezekiel xxvii. 17 is millet, or *Panicum miliaceum,* though to be fair, it must be said that Moffatt calls it "wax" and the Roman Catholic Douai version refers to it as "Rosin" or "Resin". The word *pannag,* or *panicum* is of course undoubtedly related to the Latin word "panis" meaning bread and to the French word "pain". Today millet seeds are used to feed birds in cages, while in North China today the millet is made into a kind of porridge or bread. It is a plant which perhaps produces more individual grains than any other. These are very hard and pure white and so make quite good flour. The word "miliaceum" refers to the one thousand seeds each stem is supposed to produce.

The millet is really a grass which grows about twenty four inches high; it has to be sown each year and in Italy and in Spain I have seen it grown for food, for there it takes the place of both barley and oats. I have never eaten bread made from millet but those who have tell me that it is not very agreeable and so perhaps this was the reason why in Ezekiel iv. 9 it had to be added to the wheat, barley and beans to make bread. The mixed flour had to be made like barley cakes and it had to be baked until it was brown.

Reed

I don't pretend to understand the difference between the many reeds mentioned in the Bible. Whether it is the true *Arundo donax* as for instance in Matthew xi. 7 "a reed shaken with the wind" or whether it is *Phragmites communis* as in 3 John 13, "I will not with ink and reed write unto thee". (See below.) Most Bible students seem to have difficulty in making a definite distinction and especially so as the reed on which the sponge was placed to give the vinegar drink to the Lord Jesus is supposed to have been *Sorghum vulgaris,* variety *Durra.* To complicate matters there are those who believe that all the reeds mentioned in the New Testament are *Typha latifolia,* which is the Reed Mace.

Anyway, it can be said that the *Arundo donax* is commonly called the Persian Reed and this is quite common in Palestine especially along the Jordan valley and round about the Dead Sea. This reed or grass can grow to a height of eighteen feet I am told, and at the base the diameter can easily be three inches. It bears at the top a lovely plume similar to that of the Pampas grass which is sometimes seen in gardens in Great Britain. These thick leaves are made into flutes or pipes, fishing rods, walking sticks and even into gardeners' measuring rods.

We read in 2 Kings xviii. 21 "thou trustest upon the staff of this bruised reed, even upon Egypt, on which if a man lean, it will go into his hand, and pierce it". This is certainly true of this particular reed which seems to splinter when dry and then you get knife-like sharp edges which can easily enter a man's hand. I am told that the torturers during the Inquisition used to drive sharpened portions of this hard reed in between the flesh and the toenails or the flesh and the fingernails.

There is no doubt, however, that the reed used for making pens was *Phragmites communis.* The ink used with reed pens was some kind of lamp-black, mixed with gall juice, which was carried in an ink horn suspended from the belt of the writer (Ezekiel ix. 2). This particular reed grows to a height of fifteen feet and the stems are quite woody at the base. It is found in the marshes in many different parts of Palestine.

The "pen" in 3 John xiii., "I will not with ink and pen write", refers to a reed pen, as it does in the Apocrypha—3 Maccabees iv. 20 when pens are mentioned. A special sharp knife was used to shape the pen which is referred to in Jeremiah xxxvi. 23.

When we come to the *Sorghum vulgaris,* variety *Durra,* we are in a quandary because of the tremendous difference of opinion there has been in the past. It was an annual grass, growing something like maize, which reaches a height of six feet or more in Palestine, but it can get to a height of twenty feet in some parts of the world. It bears seeds and the grains may be made into bread though this is apt to be bitter. For this reason, maybe, it is thought that Boaz gave sorghum as "parched corn" to Ruth in Ruth ii. 14.

Anyway, it has been grown since the earliest days and after the seeds have been removed the tops of this reed look broom-like and therefore they could have been used for putting the blood on the lintels and doorposts at Passover time.

As for measurements, a reed is said to be six cubits. Now a cubit was reckoned as the measurement from the elbow to the end of the middle finger—say one foot six inches. So again turning to Ezekiel—we read in Chapter xl. 5 that a "measuring reed" is six cubits long, that is nine feet.

The Hebrew word used for Reed is generally *ganah*—and only once as *agam* in Jeremiah li. 32—undoubtedly therefore a special reed which the author cannot identify. On the other hand in the New Testament the Greek word *kalamos* occurs. This could be a "cane".

Last we should consider the Kanah River in Joshua xvi. 8 and xvii. 9. This was a brook between Ephraim and

Mannasseh—probably full of reeds—for "Kanah" has the meaning of "reed or bulrushes". This reed was probably the Sweet Cane.

Rie

Curiously enough, "rye" is always spelled "rie" in the Authorized Version of the Bible. For instance in Exodus ix. 32 appear the words "But the wheat and the rie were not smitten: for they were not grown up". Unfortunately there is a little trouble in connexion with the word for in both Exodus and Isaiah it is translated "rie", but in Ezekiel the word is translated "fitches".

The rie to which the Bible refers seems undoubtedly to be *Triticum aestivum,* variety *Spelta,* while the true rye is *Secale cereale,* and it is thought that this rye was quite unknown in Egypt at that time. Egyptian experts, however, say that rye is now grown in Egypt. The rie is a grain which will grow in poor soil and even when not fully matured can easily produce a much taller stalk than wheat. The rie flour is used for bread and cakes, which are by no means as palatable as those made from wheat flour.

In all probability rie was used as one of the primary crops on new land, because it would grow on poorish soil. One gets this idea from Isaiah xxviii. 24, 25, "doth he (the plowman) open and break the clods of his ground? When he hath made plain the face thereof doth he not . . . cast in the principal wheat and the appointed barley and the rie in their place? For his God doth instruct him to discretion and doth teach him." It is as if God gives the farmer the discretion to choose which of the grain crops should be grown in accordance with the type of soil that is available.

In the Revised Version of the Bible the word "spelt" is

used instead of "rie" and in some dictionaries the translation is given as a "kind of spelt". We can now see where the variety "spelta" comes in. There seems little doubt that "spelt" is the type of wheat known in Europe as "German wheat", which has a smooth ear and not a hairy one. In the writer's view the translation of the Revised Version seems to be the right one. If this is so, then the fitches to which we referred above from Ezekiel iv. 9 were probably vetches, or a member of that family, of which the cummin forms a part.

Rush

Once again this plant hardly seems an agricultural crop but if it is what the "seven well-favoured kine" ate in Genesis xli. 2, as it may well be, then this rush *(Juncus effusus)* proved to be good food for them. It would probably grow in what is called in this country a water meadow, for Job viii. 11 says: "Can the rush grow up without mire?" I find it difficult, however, to be quite sure to which of the grass-like plants the texts refer, for botanists tell me that there are something like twenty different varieties to be found, even today, along the edges of rivers and streams in Palestine.

The correct naming of the plant is made all the more difficult because the word actually used is Egyptian in origin and not really Hebrew and that is why I imagine that some have called it "a nutritious luxuriant grass", for that is what it would seem to be from the description of the dream told to Joseph. As the rush mentioned in Isaiah ix. 14 seems to be rather tall it is wondered whether it can really be *Juncus effusus,* but at least one translator, Moffat, alters the wording to read "so he lopped off head and tail, palm-branch and rush in a single day". In that case, the palm could be tall and the rush could be quite dwarf and that would fit in with our name and description.

The function of this rush is not only to provide the leaves and stems, which are fairly soft, but the roots do a very good job of holding the soil together when it lies just below the water.

Straw

Fifteen times in the Bible *teben* is translated "straw"—and *mathben* as straw only once, in Isaiah xxv. 10. In the latter case the author feels the word should be "vegetable waste".

Most of the straw in Palestine would be barley straw. This was used for feeding cattle, horses, and donkeys —whereas the poor of the country ate the grain.

Was the straw mentioned in Exodus v. 7-18 barley straw? The author feels that it was merely the stalks of wild grasses which the Children of Israel were forced to go and gather. The straw mentioned in Genesis xxiv. 32 was surely hay and the lion in Isaiah xi. 7 probably ate hay and not straw.

Tares

Fortunately the whole situation with regard to tares seems to be a fairly simple one. It is not the tares that the British farmer knows today but what is normally called "the bearded darnel", the annual darnel grass, and sometimes an annual rye grass. The Latin name is *Lolium temulentum*, and though the seeds are much smaller than the grains of wheat, yet in the early stages of its growth the leaves and stems are so similar to wheat that it takes an expert to recognize them apart. Thus, our Lord's parable in Matthew xiii. is, as one might imagine, very real and important. If someone did sow tares among the wheat, without the farmer knowing, then in the early stages no one would suspect that anything evil had taken place for the two plants look so similar. It would be only when the plants were starting to produce their seeds that recognition would take place and then, of course, it would be very difficult to get among the plants in a wheat field in order to eliminate the bearded darnel. I am informed that in some Eastern countries where young children are allowed to work, the farmers use them for walking carefully among the wheat in order to pull up the tares the moment they are recognizable; normally, however, the two are allowed to grow together, in accordance with the Bible advice, for it is extremely difficult to pull up the tares without injuring the roots of the wheat.

The interesting thing is that there appears to be some poisonous effect when the seeds of the darnel are ground up with the grains of wheat and are used in bread. It appears that this is because of some fungus disease which grows either on or within the seed. This problem of darnel poisoning occurs again and again in the literature dealing with the subject. The fungus which apparently develops beneath the seed coats was supposed in the olden days to cause drunkenness and this characteristic gave the Biblical tares the name of "drunken darnel". There is an old wife's fable that darnel is really evil wheat and that it has, so to speak, had the curse put on it.

Once harvesting has taken place it is possible to separate the wheat grains from the darnel seeds by fanning or by sifting, thus once again the truth of the parable is brought out, that good men and evil men will be allowed to live side by side and that the final sifting will be done at the Judgment Day.

Wheat

The only problem we have to face in connexion with the use of the word "wheat" arises because of the complication of the alternative word "corn". Now "corn" in the United States of America refers to maize, known as "mealies" in South Africa. "Corn" in Great Britain usually refers to wheat, though sometimes it is used by the uninitiated to refer to any type of grain, barley, oats or wheat. Things would be made simpler if the same Hebrew word had been used in all the references to wheat. Unfortunately twelve different Hebrew words are employed altogether, one for corn in the ear, another for sheaves of wheat, yet another for ears of corn, and so on.

Having said this, it can be stated that the wheat referred to is the *Triticum aestivum* and the *Triticum compositum.* The *Triticum compositum* is the composite wheat which may have seven ears of corn, or to put it another way, seven heads on each stalk. This is undoubtedly the type mentioned in Genesis xli. 5. This wheat is seen in the Egyptian hieroglyphics and is the variety which has been called in the popular press "the mummy wheat". It was this type of wheat which was obviously grown under Joseph's instructions in the time of the Biblical Pharaoh. The harvesting would be carried out in June each year, by hand, as it is still done in Spain today.

The ordinary wheat *Triticum aestivum* has been known from the earliest times. No wonder that it is called "the staff of life", for bread is still the main food of mankind, even if in these so-called enlightened days the wheat flour is used for biscuits, cakes and packeted cake foods! Wheat growing was certainly very important in the Holy Land and in fact grain was sold to other countries as an export. We read, for instance, that Tyre was a big importer of wheat from Palestine, as we see in Ezekiel xxvii. 17.

Even today in the East there is a tendency to refer to periods of time as so many weeks after the wheat harvest or before the wheat harvest, and Reuben, in Genesis xxx. 14 actually went out in the middle of the wheat harvest in order to look for the mandrakes for his mother, Leah. The actual time of the harvesting would vary from district to district and to a certain extent in accordance with the season, but it can be said that in the earliest years and in the warmest parts it would take place towards the end of April and in the late years and in the colder regions it might occur early in June.

It always astonishes American visitors to Great Britain to discover that our fields of wheat are described as cornfields; hence the hymn, "Fair waved the golden corn In Canaan's pleasant land, When full of joy, some shining morn, Went forth the reaper-band." Scripture has it that "parched corn" was eaten again and again; it occurs for instance in Joshua v. 11 "did eat . . . parched corn", and the Hebrew word there is *qalah,* which really means "roast corn". Then you get the Hebrew word *qali* mentioned in Leviticus xxiii. 14, and 1 Samuel xvii. 17, and there the real meaning of the words "parched corn" is "roasted corn" or roasted grains of wheat.

The way this roasting is done is to get hold of the ears of wheat, bind them together in a little bundle on their stems, say one foot or so long, and then, having made a camp fire, hold the heads of corn over the flames until the chaff is largely burned off and this makes it possible to get out the grains by rubbing them through the hands and eating them. Thus the delicacy is the wheat roasted quickly over a smoky camp fire, and I am told that it is quite a favourite food. I imagine rather in a similar way that marsh-mallows are toasted and eaten in Great Britain and the United States of America.

When Joseph's brothers arrived in Egypt and he wanted to give them a dish which would please them as well as build up their health, he probably gave them "frumenty", which is really wheat boiled in milk. It is the kind of dish that children eat for breakfast today when they have porridge, only of course porridge is oatmeal. In China, even today, they have porridge made with wheat flour.

Unfortunately, some have thought that the meat offerings described in Leviticus were really offerings of meat, as we know it today, whereas the word used is in far more of a general sense, as when our Lord said: "Children, have ye any meat?" in John xxi. 5. What he was really saying was "Have you anything to eat?"—and the answer, of course, was "No". So the "meat offerings" of Leviticus and of Jeremiah xvii. 26 were really "wheat offerings" and some of the translators have purposely, therefore, made the translation "meal offerings" and others "cereal offerings".

Incidentally, there is nothing unusual at all in the cropping details given by our Lord Jesus in Matthew xiii, for it is possible to get thirty-, sixty- and hundred-fold. Wheat, of course, is still a tremendously important crop in the world, and the author was very interested, when he went to preach in Fort William and in Port Arthur in Northern Canada, to see the immense concrete granaries which had been erected to hold hundreds of thousands of tons of wheat right at the head of the Great Lakes. It was Joseph who started the practice of building granaries and of storing wheat for seven years. Even before that, householders would try to store wheat in some central place in the house.

This is made clear in 2 Samuel iv. 6: "They came thither into the middle of the house as though they would have fetched wheat." Presumably the wheat would be stored there for safety and because the centre of the house would be dry. It was possible also to store the grain in dry wells, presumably as we in Great Britain store potatoes and other root crops in clamps.

At any rate when Absalom had his insurrection in 2 Samuel xvii., verse 19 talks of the hiding of Jonathan and Ahimaaz in a dry well where they were camouflaged by having grains of wheat spread over the mouth of the well on a sheet. Those who were seeking the men must therefore have thought that it was quite a normal thing to store wheat in that way, for the men were not found, apparently.

CHAPTER 8

HOEING, DIGGING, HUMUS AND WEEDS

ONE WOULD IMAGINE that after God had made the earth as in Genesis i. 2, and had made that perfect Garden of Eden, Adam, the first gardener, had comparatively little to do. Chapter ii. 8, makes it clear, "The Lord God planted a garden", and if he planted the garden, it must have been a very wonderful and beautiful place. The trees that grew were not only beautiful but produced fruits that were good for food. There were vegetables and salads which grew and which provided food also (Genesis i. 29). Adam's job was, as Genesis ii. 15 states, "to dress the garden and to keep it". Now this word "dress" is an interesting one; it is *abad* and it means "to serve" or "look after". So you can see that the work which was enjoined was not particularly arduous but sufficent to keep Adam usefully employed during, presumably, the greater part of the day.

Then, of course, came the day when sin entered into the world. The Devil, who once had been the most beautiful angel-like being who committed the sin of pride, was thrown out (as the Lord Jesus said). He was, therefore, determined to make Adam and Eve to sin also. He tempted them, as he does men and women today, along the lines of pride, the sin which had been his downfall. One sin invariably follows another, and therefore from pride, it was covetousness and stealing and then, later on, lying. This led to man being put out of God's perfect garden and a new factor is revealed.

Adam has now to battle with weeds and has to face a

hard life of sweat and toil. We see his son, Cain, in Genesis iv. 2, labouring and working on the ground and as the result of his labours he was able to place on the altar the *peri* or fruit of the soil. This is a generic word which is used again and again in the Bible, as for instance in Genesis xxx. 2, and in Deuteronomy vii. 13, "the fruit of thy womb" and in Psalm civ. 13, "the fruit of thy works". It may be that more fruit was grown than vegetables in those days, for throughout the Bible, greater importance is attached to trees than flowers or herbs. Remember that, "a river went out of Eden to water the garden" (Genesis ii. 10).

Whether or not vegetables were grown in these early days, they certainly were grown later on, for when the Israelites were punished for their sin by having to wander in the wilderness, they longed for the delicious vegetables they used to grow in Egypt, irrigated by the waters of the Nile.

There are old-fashioned gardeners who say: "Never let the weeds see a Sunday." This means, of course, that you must hoe the garden during the week so that by Saturday evening there will not be a weed left in the garden to see the Sunday that follows. As a result, of course, you will never have any weeds. There is something about this in the Word of God, for in Isaiah vii. 25 it says: "And on all hills that shall be digged with the mattock, there shall not come thither the fear of briars and thorns."

In the Revised Version the word "digged" is translated "hoed" and so it is obvious that the Prophet was saying that if you hoe regularly, not only will there not be any weeds—but, better still, you will not even have the fear of weeds! Bible students will note that a similar word is found in 1 Chronicles xii. 33, when it talks about the soldiers of Zebulon which "could keep rank", and the term "keep rank" had the same Hebrew origin as "hoe". Today we grow our vegetable or salad crops in their straight lines or ranks and the gardener who hoes regularly to keep these ranks in perfect order finds that it makes all the difference to the success of vegetable gardening.

When the author was a Colonel in the army, responsible for the food production of the units, he was able on many occasions to get the men in lines with hoes in their hands working up the rows of vegetables on quite large acreages. Here the soldiers did their hoeing almost by "numbers" and as they walked forward, hoeing as they went, they were encouraged to keep up with one another and so they "kept rank". It would be useful indeed to have such a gang today in many gardens!

Regular hoeing is tremendously important with the normal cultivation of crops. It need not be done deeply and in fact the disturbance of the top half inch of soil will do. The idea is to kill the weed seedlings just as they start to germinate and not to let them grow to any height at all. Weeds, of course, compete with the plants one is trying to grow, for food and water and they may easily grow as strong as our Lord suggests in the parable of the sower (Matthew xiii. 7) "And some fell among thorns; and the thorns sprung up and choked them". The real translation here should be "thistles"; in fact, even today, we have a cultivated type of thistle known as *Acanthus.*

It is obvious that the seed here fell on an area of ground that was full of thistle seed; and, because the baby thistles were not kept under when they started to spring up, they completely smothered the good wheat that was trying to grow. It is indeed the perfect picture of what happens in this world, which our Lord describes in verse 38 of that same chapter.

The people of Palestine in the days of the prophets knew little of the deep cultivation which farmers and gardeners think is necessary today. Their implements were quite light and when sowing seeds they merely scratched the surface of the ground, rather than ploughing deeply as we do today. There are those who have wondered whether we have not erred in trying to cultivate too deeply and certainly at the Missionary Horticultural College we are preparing the soil far, far more shallowly than we did years ago. The tendency, too, is to apply the fine organic matter on the surface of the ground and to work it in, rather than to bury the dung or compost deeply. There was no deep burial of manure in the days of the Bible.

Humus is the Hebrew word *aphar.* It was because the original translators of the Bible knew nothing of the living soil that they translated the word "dust". Dust, of course, is dead—and God has condemned those who will not obey him, Deuteronomy xxviii. 15—while verse 24 says that disobediance will result in "the land turning to powder"; the punishment for not giving back to the soil what was taken out by the plants.

God in creating the earth chose to take plants, animals and man from the humus of the soil. It was when Adam sinned by coveting, lying and stealing, that he was turned out of the everlasting garden when he himself lost his own everlasting life. It was then that God told him that he would die and be buried in the soil for, said Almighty God, "Humus thou art and to humus thou shalt return".

Thus God's plan for agriculture and horticulture was

revealed. The animals in the forest when they die go back into the soil, so do the bees, wasps, flies and birds. The leaves fall down in the autumn and form a carpet and the worms come up and pull leaves into the soil. These leaves have had some excreta dropped on them and this acts as the activator when in the soil. Thus natural compost is made.

The name Adam means "from the soil", the term *Adamah* is organic husbandry or the correct care of the soil. When there is plenty of humus in the soil, then the land will produce excellent crops with a certain immunity from diseases and pests. Humus feeds the millions of living organisms in the soil. It assists the hormones, the enzymes, the function of the mycorrhiya and so on.

Humus is the "blood" of the soil. It stores water and plant foods. It keeps the soil open, ensures the right aeration and is an insulation against cold and heat.

We are apt to have an idea that we of today are the enlightened peoples and that the patriarchs in fact knew very little about the cultivation of the soil. There are, however, those like the author, who believe that men like Isaac were very skilled in their management of the soil and, maybe, this was God-taught. Anyway, in Genesis xxvi. 12, we read that as a result of the sowing of the seed, the return was an hundredfold, and the report is followed by those words, "The Lord blessed him and the man waxed great and went forward." Here indeed is an example of temporal blessings and of very definite earthly prosperity.

Whether or not his work was to be an example to the Philistines, the fact is that he did the seed sowing in their land where he had gone during the famine. And because of his great success as a farmer, we read: "the Philistines envied him", Genesis xxvi. 14. We are not told how Isaac manured the land, or fed it, but there is very little doubt that it must have been along the lines of the suggestions made in Chapter 10.

We imagine that the work of soil preparation was done with primitive ploughs, for there are a number of references to plough yokes and ploughs, as in Genesis xxvii. 40 and Judges xiv. 18, while in order to emphasize the sad results that occur when a Christian marries an unconverted person, we have the definite order, "Be ye not unequally yoked together with unbelievers" (2 Corinthians vi. 14). This, of course, refers in the first place to the definite injunction of Deuteronomy xxii. 10, to the farmers of those days, never to yoke an ox and an ass together.

One of the ways in which we modern gardeners obtain a sufficiently fine tilth in which to sow the seed, is by the use of the rake. This is moved backwards and forwards evenly and the clods, if they are in the right condition, are thus broken down. One often has to wait for a rain before the lumps or clods of soil will disintegrate. This bringing down or raking down to a tilth is the Hebrew word *chathath* which is found in Jeremiah xiv. 4. It was impossible to prepare the soil for seed sowing "for there was no rain". Many is the time the market gardener waits for a shower of rain before he can harrow a particular piece of land and the harrow to the commercial grower is what the rake is to the private gardener. We read in Job xxix. 23 "They waited for me as for the rain."

One of the big punishments threatened to the people was the withholding of rain; for not only was this needed to nourish the crops, but as we have just said, it was tremendously important in the preparation of the ground for the acceptance of the seed. 1 Kings viii. 35:

"there is no rain, because they have sinned". On the other hand, rain was promised in answer to prayer, Zechariah x. 1: "Ask ye of the Lord rain . . . so the Lord shall . . . give them showers of rain."

This raking or harrowing was obviously quite a common practice in the days of David, for one of the terrors to which the Ammonites were subjected was to be killed by heavy iron harrows being dragged over them (2 Samuel xii. 31). Whether this laying of these immoral Ammonites on the ground and harrowing them, so to speak, into the soil, was part of the manurial plan for the land, I do not know. The word "harrow" used here, however, is *charits* which really means "pikes", the kind of implement used by the pikemen in the days of Elizabeth I.

On the other hand, just below this word "harrow" in Young's Analytical Concordance, there is the second Hebrew word *sadad* which also means to harrow, or better still, to level. Thus the raking done by the gardener is not only to break down the clods, but it is also to leave the soil level, so that all the rain will reach the ground equally well and will not collect as puddles in the slightly lower portions. A levelling of the soil makes for more even germination and a general evening up of the growth of the plants in that part of the garden. Job xxxix. 10 says: "Will he harrow the valleys after thee?" This refers to that strange beast of the Authorized Version, the unicorn. It is better translated as the wild ox; and the picture here is of the farm labourer leading the ox and the ox itself pulling the heavy iron harrow behind it.

Cloddy land can be a nuisance. Seed can be buried too deeply if it is covered by great lumps of soil and this is presumably what is meant in Joel i. 17, "The seed is rotten under their clods". These clods were presumably hard and baked, whereas when we turn to Job xxi. 33, we read: "The clods of the valley shall be sweet unto him". The idea here is that the clods would have been sweetened by the rain and would therefore break down easily, with the harrow. Furthermore, to spiritualize the point, there shall be nothing frightening or sad about the grave; in fact one could say that this nice tilthy soil would be a great joy and that his body would be forming good humus while his soul, of course, returned to his Maker (Ecclesiastes xii. 7).

The gardener dislikes the condition outlined in Job xxxviii. 38, "When the dust groweth into hardness and the clods cleave fast together". Here you get the picture of a soil that is lacking in humus and so grown together like baked clay and the clods, as a result, cannot be broken up and a good fine tilth cannot in consequence be prepared.

Ferrar Fenton who translated the Bible into modern English from the original Hebrew often used the Hebrew rhythms of the prophets. His translation of Isaiah xxviii. 23 onwards reads: "Give ear to my voice—hear my words with attention—the plowman ploughs daily and harrows his land for the grain; when level its surface, sows he not? . . . for God hath instructed and taught him reflection."

We learn so much empirically and much that we do today in the garden is done because we have learnt the best method in a hard school. The gardener must be observant and he must learn day by day by what he sees. It is wonderful to realize that God can cause us to reflect, to think back, and that God can instruct the gardener, and has instructed the gardener in the past. In this particular passage we learn that the gardener must know the right depths to sow his seed and whether to broadcast it or sow it in straight lines or drills.

Not only must we gardeners know about the preparation of the soil and the sowing of the seed but we must be able to harvest our crops at the right time and learn the right ways of doing this work also. The passage goes on to deal with the way the various crops should be harvested, followed by the words, Isaiah xxviii. 29: "This also comes from the Lord of all powers, in purpose the Wondrous, the Great in effect." The God who teaches us how to sow, instructs us equally well on the method of gathering the crop for our use.

It is obvious that God expects his people to be good craftsmen. For instance, it is no good gardening unless one takes infinite trouble. Thus when digging, it was expected that every single little root of a thistle should be removed during the operation: "all hills that shall be digged with the mattock, there shall not come thither the fear of briers and thorns" (Isaiah vii. 25). One could almost believe that the term "mattock" here should really be translated "fork", for it is as one forks over a piece of land carefully that the roots of perennial weeds can be picked out. Actually the word is *mader* which should be translated "rake". If these rakes had very long teeth, then they could disturb the ground in a similar manner as the fork and so one could get rid of the roots.

There are unfortunately many people today who are called "fair weather gardeners". They are never seen on their allotments or gardens until Good Friday and the moment there is any sign of wintry weather in the late Autumn they creep back into their homes again! We are warned against such laziness when we read in Proverbs xx. 4: "The sluggard will not plough by reason of the cold; therefore shall he beg in harvest, and have nothing."

The digging of the land in the Autumn and leaving it rough, is part of the process of ensuring that the clods of the soil do break down. The frosts and cold winds act on the lumps of earth and cause them to become friable almost automatically. Land that has been weathered, as we gardeners say, can quickly be forked and raked down to a fine tilth in the Spring. Land on the other hand that is beaten down by the wintry rains, then is frozen stiff, may easily set hard like cement and be very difficult to prepare in the Spring.

Before we leave the whole question of the cultivation of the soil, we must turn to the New Testament where we find in Luke xiii. 8 that dung was to be cast around the tree as a mulch and it is the only place, as far as I know, where the suggestion of organic mulching is given. The translation "and dung it" does not give the true picture, for the Greek words are *ballo koprian* which literally means to throw dung "on top". The word here "to dig" is *skapto* which may mean "to delve" and can mean "to hollow out" and as the text says:—"till I shall dig about it". It seems to the author to infer that a very shallow trench was hollowed or scooped out and the dung was then put into this as a mulch. It all probability some water would have been poured on top of the ground, which being in the shallow trench would not blow away, as it might do if just put on the surface of the soil, in a hot country like Palestine.

Apart from the reference in Luke xvi. 3, where the unjust steward says that he is not a skilled gardener and that he cannot dig, most of our references in the Bible as regards digging refer to the preparing of wells. Genesis xxvi. 19 for instance: "Isaac's servants digged the well"; There is the digging when preparing for a winepress as in Matthew xxi. 33 and Mark xii. 1, while the word *adar* which appears in Isaiah v. 6 and vii. 25 really seems to mean "to be set in order" or "to be arranged neatly". We certainly can say this, for the perfect digging of a good gardener does indeed leave the land beautifully arranged and "set in order".

Those who have advocated no digging (and at least two books have been written on the subject) have contended that God's plan must be to sow the seed on the surface of the ground (in the same way as weed seeds are normally distributed) and it is argued that they grow magnificently as a result. Here, of course, it is the survival of the fittest, for every plant that seeds itself naturally, invariably seems to have thousands of seeds to distribute. Fortunately, not all these germinate, despite what gardeners may believe! This survival of the fittest does, of course, ensure that the weeds go on year after year, whether they be dandelions or docks, or whether they are the daisies in the lawn or the poppy in the farmer's field.

The Egyptian gardeners must have been surface-soil sowers, and presumably the beautiful fine tilth produced by the overflowing river on the banks was in a wonderful condition for the sowing of seeds. It must have been something like the special compost prepared by the greenhouse grower for the raising of precious seedlings. As the animals came down to drink, they would naturally leave their droppings and urea behind, and in all probability therefore the lovely silty loam on the sides of the Nile was not only in a perfect condition for seed sowing but organically fertilized also. The casting of the seed, therefore, upon the waters, may refer to the rice growing, or it might equally well be shown to refer to the sowing on the land that had been covered with the waters. One can turn from Ecclesiastes xi. 1 to Isaiah xxxii. 20 where I think we see the picture in its true perspective—"and the sowers by all streams shall be happy, who work with the ox and with the ass" (Farrer Fenton). Once again, we see the tilth and the natural manure from the ox and the ass.

As one who has taught hundreds of women to garden, I may say that this is indeed a reference to women gardeners, or certainly women farmers, for this Ode No. 50, as Farrer Fenton puts it, which commences at verse 9 of Isaiah xxxii and ends at verse 20, is headed "The sin of female luxury" and the "idle girls" in verse 11 are told to "mourn on the fields".

It would be wrong not to include in this reference to the surface sowing of seeds, the parable of the sower in Matthew xiii. It is the word *speiro* which means "to sow the seed" and is repeated again in Mark iv. and Luke viii. It has been suggested that because our Lord said that some of them fell by the wayside in stony places and among thorns, it was obvious that the seed was being sown on the surface of the ground, because we read, "others fell on good ground". I have purposely used the word "on to", for the translation "into" is not absolutely correct and therefore the version given in Mark and Luke is right when rendered "on". Those who argue this way, must find the answer to the problem why birds should prefer to seek for seed on the hard path beside the cultivated land and leave that which fell on the prepared earth.

It would seem, therefore, that the land had been harrowed and so prepared or, as we gardeners would put it, had been raked down to a fine tilth. Then the seed was sown on the fully prepared soil and was raked in lightly to cover it and to protect it from the ravages of the birds. Thus it grew and brought forth, some thirty-fold, some sixty-fold and some a hundredfold.

Nettles

Nettles are a very common and persistent weed in Great Britain, but whether the Hebrew words *charuh* and *qimmosh* or *qimosh* are true nettles I very much doubt. *Charuh* is really a thorny shrub and *qimmosh* a thorn. The word "nettle" however appears eight times in the Authorized Version.

In Job xxx. 7 we find "They were gathered together under the nettles" so this translation of the word *charuh* hardly fits. On the other hand, "Nettles shall possess them", in Hosea ix. 6 seems good, but the Hebrew word is *qimmosh.*

The Job nettles are probably the tall perennial *Acanthus syriacus* which is known in Palestine where it grows freely. This is a plant that could give protection.

In Isaiah xxxiv. 13 we have nettles again, probably *Urtica pilutifera,* and they could be the ones known in Great Britain today. As they are often seen growing around ruins they do fit into the picture of desolation.

Rolling Thing

Is this Hebrew word *gulgal* found in Isaiah xvii. 13 "a wheel" or a roller? This plant today is the Resurrection Plant (*Anastatica hierochuntica*)—sometimes called the Tumbleweed. In Jericho I found it was called the Rose of Jericho. The plant loses its leaves after flowering and the stems curl round and form a hollow ball. This then rolls away, blown by the wind, and as it travels it drops its seeds here and there. It is possible in Palestine to see dozens of these "Rolling Things" in the fields or roads and they make quite a noise as they travel along.

The Professor at the Hebrew University however feels that *gulgal* is really *Gundelia tournefortu,* a type of thistle, which behaves like the *Anastatica.*

Thistles

In the Old Testament, *dardar* and *choach* are both translated "thistle" and in the New Testament the Greek word *tribolos* is "thistle" also.

It all starts in Genesis iii. 18 when thistles are promised to Adam after he sinned. The thistle converses with a cedar in 2 Kings xiv. 9 and 2 Chronicles xxv. 18. But I found over one hundred species of thistles in Palestine. It is a very common weed there now and I expect it was in Biblical days.

In Matthew vii. 16 the word "thistles" should be "briars" undoubtedly, as it rightly is in Hebrews vi. 7 "thorns and briars".

Moffatt, who invariably goes his own way, translates the Hebrew word *dardar* as "thistle" each time and when the word *choach* occurs he translates it "thorns"—and I cannot help feeling that he is right.

But in Isaiah xxxiv. 13 the word *sir* occurs and in our Authorised Version we read "brambles". I feel that this is the Palestinian thistle *Solymus maculatus*—a spotted golden kind. Other thistles in Palestine are the Lady Thistle, the Dwarf Thistle, the Star Thistle and so on.

Referring back to the parable in 2 Kings xiv. 9 above, I think that the word here should read "thorn bush" or "bramble" and not "thistle". But as one can have a six foot thistle—*Onopordon arabicum*—was this the kingly thistle that felt it could raise its voice to a Cedar?

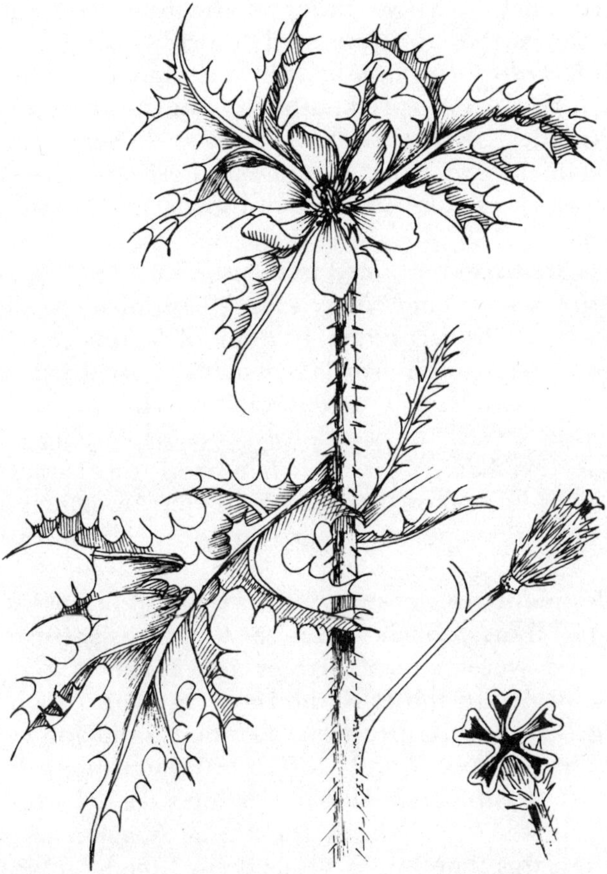

THISTLE. (Devil's Fig). Argemone mexicana.

Thorns

This is perhaps the most difficult name to pin down. It is mentioned fifty-two times in the Bible and the Hebrew words that are translated "thorn" include *atad* (bramble), *chedeq* (briar), *choach* (thicket) *naatsuts* (thorn), *sallonim* (briars), *tsen* (prickly-thorn) *qots* (thorns) *qimmashon* (nettles) *shayith* (thorns) all in the Old Testament; while in the New Testament there are Greek *akantha* (briar); *akanthinos* (made of thorns) and *skolops* a sharp stake.

Thorns are first mentioned in Genesis iii. 18 owing to Adam's great sins. Then we read in 2 Chronicles xxxiii 11; Job xli. 2; Proverbs xxvi. 9; Song of Solomon ii. 2 and Hosea ix. 6—the word "choach". The word is translated "brambles", however, in Proverbs.

In Hebrews vi. 8 the word *tribolos* (a triple-pointed plant) the text being "that which beareth thorns and briars", must be one of the thistles, in fact *Acanthus* is a British thistle-like perennial grown in a Herbaceous border.

In Palestine there were numerous types of brambles, briars and thorny bushes. These could easily form thickets and were natural hedges, see Isaiah v. 6 or Proverbs xv. 19. In Job 1, 10 the Devil complains about the hedge around the property. Continuing in Job we come to chapter xxx. 7 and Moffatt translates the word "nettles" as "scrub". The Douai Version calls the word "briars". In Zephaniah ii. 9, the Douai Version uses thorns—but this time Moffatt calls them "weeds". Once again the author wonders whether the answer is not the Acanthus.

There is another problem in the Song of Solomon, for it is difficult to believe that lilies grew among the thorns, brambles and briars as in Song of Solomon ii. 2.

Were the lilies among nettles—among Acanthus—or what? There had to be, of course, the vital contrast between his lovely girl and the other less beautiful daughters. Is this the desired contrast in nettles and lilies? I wonder.

CHAPTER 9

THE DISEASES AND PESTS IN THE BIBLE

If Adam had never sinned and if the future generation had kept faithful and true, presumably we gardeners would not be worried day by day by the ravages of pests and diseases! Genesis iii. 17 says: "cursed is the ground for thy sake; in sorrow shalt thou eat of it all the days of thy life . . ." And in verse 23 "Therefore the Lord God sent him forth from the garden of Eden, to till the ground from whence he was taken." The fact is that right from the earliest times in Scripture, details of pests occur, and diseases too for that matter, referred to usually as "blasting and mildew".

We will deal with the diseases and pests in alphabetical order, despite the fact that in the case of some there will be but little information to give, while in the case of others like the locust, pages could be written.

Ants

Ants are mentioned in Proverbs vi. 6-8 and in Proverbs xxx. 25. In both passages they are praised for their industry, and those who have studied these creatures know what hard workers they are, and how they can give instructions to one another by means of their antennae. They appear to be abundant in Palestine, and have undoubtedly been so since the earliest of days. They are a nuisance to the gardener, because they make their nests underneath plants, whose roots in consequence will dry up. In addition they have the nasty habit of transferring an aphis (greenfly) from one plant to another, because they like the honey-dew which is exuded from the cornicles of these creatures. They will thus attempt to establish new colonies of these insects on other trees and plants. This of course is clever of them, but it does cause the modern gardener many a headache.

Beetles

Beetles are mentioned only once, in Leviticus xi. 22. It is the Hebrew word *Chargol* and many think that this is wrongly translated "beetle", for no true beetle has "legs above its feet to leap withal". In all probability therefore, the original writer was referring to another species of locust, in addition to those already mentioned.

This does not mean to say that there are not any beetles in Palestine, and in fact some have said that they have found over four hundred species in that country. There are many beetles in Great Britain which are pests, the Shot Hole Borer Beetle for instance, the Raspberry Beetle, the Colorado Beetle, the Click Beetle, the Chafer Beetle and so on. There are on the other hand many species which do good rather than harm.

It would only be fair to mention that the word "beam" in Habakkuk ii. 11 should probably be rendered "beetle". It is the Hebrew word *kaphis,* which may be rendered "the Scarab Beetle", which was of course known to be the sacred beetle of Egypt. In the Latin Vulgate translation the word "beetle" is in fact used. It would seem also from the context that the word "beetle" is more likely, for many are the beetles that live in timber, but "beam in timber" does not seem to make sense.

Cochineal Worm

Though the actual word "cochineal" is not mentioned in the Bible, yet the Hebrew *Tola* should really be translated "crimson worm" in Isaiah i. 18. The sense of the text, however, makes it unnecessary for the word "worm" to appear, because it is the colour that is important, and not the insect from which it comes. The text is, "Come now, and let us reason together, saith the Lord; though your sins be as scarlet, they shall be as white as snow; though they be red like crimson, they shall be as wool."

It is from the female that the crimson dye is obtained. It attaches itself to the plant or tree, and remains there immovable during the rest of its life. After impregnation by the male, the female may resemble a convex mass without the least appearance of head or feet. This insect produces the most brilliant dye, as well as the basis of the most useful kinds of cement. It is one of the insects to which we human beings are particularly indebted.

It was known to the Phoenicians before the time of Moses, and is found in great abundance on a species of evergreen oak.

Flies

While there are a number of types of flies that cause a good deal of harm in Great Britain, like for instance the Onion Fly, Carrot Fly, Narcissus Fly, it does not seem that the flies mentioned in Scripture are those which give trouble to horticultural crops. The *Arob* or *Oreb* of Exodus viii. 21 and of Psalm lxxviii. 45 are the flies which plagued the Egyptians when Pharaoh was unwilling to let God's people go. In all probability, this *arob* refers to the mosquito though, of course, the common fly will carry the infection of the poisoned eye from man to man.

The fly in Ecclesiastes x. 1 is probably the gadfly, which is well known to torment horses on the Nile, while the fly of Isaiah vii. 18 may be the gadfly or more reasonably the tsetse fly which causes death to animals in Egypt and Africa.

Grasshopper

Though Judges vi. 5 mentions grasshopper, as does Leviticus xi. 22, it is likely that the Hebrew words *arbeh* and *chagab* really mean "small locust," and the locust certainly was a terrible pest.

Hornet

This is found in Exodus xxiii. 28 where God says "I will send hornets before thee, which shall drive out the Hivite . . ." The Hebrew word is *Tsirah* and its Latin name *Vespa crabro.* It is a dreadful waspish insect, fierce and formidable. It is usually about an inch long, with a yellow front and an abdomen with blackish brown rings. It lives in holes in trees, walls, and buildings, and it feeds on the nectar of flowers and fruits.

It will attack hive bees, and suck the honey and moisture from their bodies. It also will attack horses, cattle and human beings.

There is a Valley in Palestine known as the Place of Hornets, see Joshua xv. 33, the word *Zoreah.*

Locust

The locust is undoubtedly the most serious pest mentioned in the Bible. Sometimes, it is true, it is called by other names, like Canker Worm, Palmer Worm, and Caterpillar. There are about forty species in Palestine, all of which can do an immense amount of harm, and all of which, curiously enough, are used for food by the natives. The migratory locusts possess the most formidable powers of destruction. They turn fertile fields and gardens into deserts. They fly over like clouds, and literally darken the sky. Years ago in South Africa it was possible for an area of two thousand square miles to be covered by a cloud of these creatures.

Because this pest is so important, nine separate Hebrew words are used to express the locust. (a) *Solam* found in Leviticus xi. 22, called "the bald locust" because it had a smooth head. This frequents the rocks and rocky land in Palestine today (b) *Chargol* found in the same verse which has been wrongly translated "beetle". (c) *Chagab* translated "grasshopper" but probably a small species of locust. (d) *Gazam* as in Joel i. 4 translated "palmerworm" though it is probably "hopper", that is, the immature locust, before it develops its wings. (e) *Yeleq* of Nahum iii. 15, Jeremiah li. 14 translated sometimes "canker-worm" and on other occasions "caterpillar". Once again, this appears to be the young locust, for the literal translation of *yeleq* is "licker of the grass". (f) *Tselatsal* found in Deuteronomy xxviii. 42 "All thy trees . . . shall the locust consume". Actually the real translation is "cymbal" or "tinkler," a name given to the locust because of the noise of its wings. (g) *Geb* or *gob,* as in Isaiah xxxiii. 4 "as the running to and fro of locusts" translated in Amos vii. 1 as "grasshopper". The real meaning of the word is "green worm", and presumably this refers to the crawling stage of the locust. (h) *Chasil* usually translated "caterpillar", but always inclued with the locust; the actual meaning of the Hebrew word is "consumer" (Psalm lxxviii, 46), and once again it is undoubtedly the locust in its early stage.

(i) The word *Arbeh* is found for locust in 1 Kings viii. 37—2 Chronicles vi. 28—Psalm lxxviii. 46—Joel i. 4—and Joel ii. 25.

Locusts are extremely difficult to control, even with modern methods, and literally tens of thousands of pounds are being spent today in various parts of Africa to keep down this astonishingly virile pest. The locust bands certainly prove the power of God to punish a wicked world at any time. The flying squadrons appear in the Spring, millions upon millions deposit their innumerable eggs, and in eight weeks billions of crawlers begin to creep. Soon these change into small locusts, and they start to grow, march and fly, devouring every green thing they come across, and leaving literally nothing behind. Joel i. 7 makes this clear, when he says "He hath laid my vine waste, and barked my fig tree: he hath made it clean bare, and cast it away; the branches thereof are made white." No wonder the prophet goes on in verses 15 and 16, "Alas for the day! . . . a destruction from the Almighty shall it come. Is not the meat cut off before our eyes. . . ?"

Vines will disappear, so will the corn, grass, weeds, flowers, and in fact all plant life, for miles and miles in the path of the locust cloud. No wonder the prophet says in Joel ii. 3 "behind them a desolate wilderness; yea, and nothing shall escape them." Furthermore, his description of how the locusts move, is so true: "they shall climb the wall like men of war; and they shall march everyone on his ways, and they shall not break their ranks." If there is a house in the way they will quite happily climb all over it in order to continue their onward march. We read of this in Joel ii. 9 where the prophet says "they shall climb upon the houses; they shall enter in at the windows like a thief."

The prophet Nahum in chapter iii. 17 mentions something about the locust which is absolutely true, though scientists in the early Victorian days doubted the statement. The verse says they shall "camp in the hedges in the cold days, but when the sun ariseth they flee away". This is exactly what happens, for they will so to speak, temporarily camp at the bottoms of the hedges and walls, like a swarm of bees, and the moment the sun rises and it gets warm the following day, off they go again on their dreadful consuming march.

The people of the East are certainly frightened by the locust, and this is no doubt the reason why Pharaoh's servants in Exodus x, tried to persuade the King to let the Children of Israel go. When Pharaoh refused and the locust did come, we read "They covered the face of the whole earth, so that the land was darkened and they ate every herb of the land . . . and there remained not any green thing in the trees or . . . in the fields."

It is very difficult to paint the true picture with regard to the locust, because no one who has not seen the swarms or clouds of these creatures can really believe that it is possible. Sufficient to say that agricultural experts of north-west Africa have stated that over thousands of acres the locusts or their larvae may be an inch deep on the ground. Fortunately the swarms may sometimes be carried out to sea by the wind, and captains of ships report that they have landed on deck at least a thousand miles away from land.

That God can allow destruction by the locust, and does in fact do it, is seen in Psalm lxxviii. 46, "He gave also . . . their labour unto the locust", or in Psalm cv. 34 "He spake, and the locusts came". Proverbs xxx. 27 states "The locusts have no king, yet they go forth all of them by bands", while perhaps the most famous text of all is found in Joel ii. 25 "I will restore to you the years that the locust hath eaten". This verse is a tremendous comfort to those who are converted to Christ late in life.

We now come to the locust used as food. The permission was given in Leviticus xi. 22 "These of them ye may eat; the locust . . . " There is no doubt that the locust referred to in Matthew iii. 4 and Mark i. 6 is the true locust, and not the locust bean as some people have suggested. John the Baptist ate locusts and wild honey, just as the Arabs eat them today. In fact, I am told there are shops in parts of Iraq, Iran and Transjordania where locusts are sold by measure in shops. It is true that these seem to be eaten only by the very poorest of people, but then that is exactly how John the Baptist lived. He would get the wild honey from the cracks in the rocks where he preached and lived.

Moth

The Hebrew word *Ash* is translated "moth," though it could be translated "worm." In the New Testament the Greek work is *Ses* as in Matthew vi. 19 and Luke xii. 33 "where moth and rust doth corrupt". Though there are many moths that give trouble in Great Britain to fruit trees and bushes, the Raspberry moth, for instance, the Magpie moth, the Codling moth, the moths of the Bible appear to be entirely those which damage clothes. Once again, it is the larvae of the moth that do the damage in the garden and in the house, by nibbling the fruit or plant, or in the case of the house-moth the wool of the clothes, curtains and carpets. It is said that over two hundred species of Lepidoptera (the moth and butterfly group) are found in the Holy Land.

Mouse

Mice are mentioned six times, the Hebrew word being *Akbar.* The first reference is in Leviticus xi. 29 in the list of the creatures that cannot be eaten. It again occurs several times in 1 Samuel vi. where five model golden mice are made by the Philistines as a trespass offering.

These mice were probably similar to the white-bellied field mice found in Great Britain, which do damage to agricultural and horticultural crops. In fact in 1 Samuel vi. 5 they are referred to as "mice that mar the land". Mice of course eat corn, they can damage root crops, they burrow into root clamps and stores and generally consume food which may be needed by human beings. They often take the peas and beans soon after they are sown. In Great Britain they are trapped by marrow seeds being used as bait in the backbreaking traps that can be bought from the ironmonger.

Snails

The Snail is mentioned twice, and in both cases there is a question as to whether the translation is absolutely correct. The word *Chomet* occurs in Leviticus xi. 30 being translated "snail", but it may easily be another species of lizard, despite the fact that the lizard already appears in the verse. In Psalm lviii. 8 the Hebrew word "Shablul" translated "snail" may easily be "festering sore". The verse says: "As a snail which melteth, let every one of them pass away," and it could be, "as a festering sore disappears, let every one of them pass away". Those, however, who believe that it is truly a snail, suggest that the melting refers to the slime which the snail leaves behind as it moves along.

Snails of course in this country are a regular pest, but

not so much so as slugs. They are night marauders, and usualy eat anything fresh, green and succulent. There are the large Garden Snail, and the Banded Snail which is more injurious to farm crops than to garden crops on the whole.

Fungus diseases

Mildew

The word "mildew" in the Bible, invariably occurs with the word "blasting". It is the Hebrew word *Yeraqon* which may be translated "paleness", and is found in Deuteronomy xxviii. 22, 1 Kings viii. 37, 2 Chronicles vi. 28, Amos iv. 9—Haggai ii. 17, the latter reference being, "I smote you with blasting and with mildew".

The paleness of the leaf may of course be caused by the mildew fungi, living underneath. Many are the mildews in this country which give such a pale effect. It is the sort of generic term which the old-fashioned gardener in this country commonly calls "blight". In the case of the dream of Pharaoh, for instance, the attack on the ears of corn in Genesis xli. 23 was undoubtedly a type of mildew.

Again and again it appears from the Word of God that a blasting wind brings the blight or mildew with it. Hence the use again and again of blasting and mildew together. Ezekiel seems to understand this, because he refers to the withering which can be done by the east wind in chapter xvii. 10.

Many of the prophecies of the Old Testament seem to be bound up with a wind and with mildew. After all, one of the ways in which God could punish a disobedient people was to injure their crops, and the earliest reference to mildew deals with this very point: "The

Lord shall smite thee . . . with blasting and with mildew". Thus it can be said that the people were warned that the Lord who controlled the winds could make these particular blasting gales to come, and bring the fungus diseases with them.

CHAPTER 10

MANURING AND IRRIGATION IN THE BIBLE

THERE WERE AT least three cities in Israel named Madmenah: (1) A town of Benjamin near Jerusalem, (2) A town near Ziklag in Judah and (3) A town on the borders of Moab.

Madmenah is translated "dunghill" in the Authorized Version, but it is obviously not dung but the compost heap, which produces the right organic matter which will provide the necessary humus in the soil and *this* is vital. Humus has been described as the "blood" of the soil, thus no humus, and the soil is dead.

It was in Genesis iii. 19 that God said to Adam "Humus thou art and to humus thou shalt return", thus laying down the basic principle of the complete circle, i.e. the passing back into the soil of that which was taken out for God added "for out of it wast thou taken".

So the Israelites knew the importance of rotting down the straw, the vegetable waste, the refuse from all the towns, and the composting was done with liquid manure—and Isaiah xxv. 10 suggests that the tougher material put on the heaps was broken down by chariot wheels being run over it. The Madmenah towns must have specialised in compost production.

The Lord Jesus obviously knew all about composting when He referred in Luke iv. 35 to salt as being useless as an activator when it had deliquesced, for then "It had lost its savour," or power to cause the vegetable waste to turn into compost.

As has already been said, it is impossible in the East to disassociate gardens and water. In fact, the dominant features of the gardens of the Pharaohs were undoubtedly (a) shade and (b) water. The very oldest garden scenes show gardeners watering from water skins or even from pails suspended from a wooden yoke. The most beautiful and interesting illustration of ancient gardens is that depicted in *Manners and Customs of the Ancient Egyptians* by Sir J. G. Wilkinson. It is the garden of a General of Amenophis III, who lived in 1500 B.C. at Thebes. In the plan there can be seen, in the greatest of detail, the vines, the lotus flowers, and the groves of trees. Four quite large water gardens are included, and these seem to be surrounded by brass pots. They may easily have been reservoirs of water used for irrigation purposes.

Generally, the kitchen gardens were laid out in a chessboard arrangement, where the cucumbers, melons, leeks and onions could be grown. In between the squares which formed the chessboard the water could run. This fact is referred to in Deuteronomy xi. 10. The gardener with a deft movement of his foot could, as it were, undam the temporary mud wall forming the channel, so as to let the water flow into the next furrow or square. Many of us have had experience of this kind of thing when playing as children on the sands of the seashore.

It seems that in the very earliest times Palestine was quite a network of aqueducts, and that was probably the reason why Lot chose the plain of Jordan, because "it was well watered everywhere" (Genesis xiii. 10). Balaam in his prophecy in Numbers xxiv. 7 says: "as gardens by the river's side . . . as cedar trees beside the waters. He shall pour the water out of his buckets, and his seed shall be sown in many waters". The reference to the buckets here is to the old-fashioned well-wheel which is still to be seen in Spain today. To this wheel are affixed buckets, which, when they are motivated by an ox or donkey going round and round, draw up the water from the well.

Some people have not quite understood the reference in Ecclesiastes xi. 1 to "waters". It refers to the water furrows dividing the normal chessboard pattern. In fact, the plural word is used to show that the main stream was divided into many narrow irrigation "canals" in order that the various beds in the garden might be supplied.

Often in the Bible, the word "rivers" is used, when the term should really read as "furrows" or "trenches". Take, for instance, David's well-known statement in Psalm i. 3, "and he shall be like a tree planted by the rivers of water, that bringeth forth his fruit in his season". He is speaking, of course, of the righteous man. The real translation should be, "He shall be like a tree planted by the side of the trench", or "by the side of the water furrow".

It would seem as if the Bible was laying emphasis on the difference between living waters, waters which flowed from a source and were constantly being aerated and renewed, and water stored in large tanks and cisterns which was available as the result of trapping and collecting the water during the rainy season. Though there had been a tendency in this country to suggest that the stored rainwater was particularly useful for greenhouse plants, yet the Bible seems to insist that the greater fertility comes from the "living water".

Jeremiah, for instance, says in chapter ii. 13, "they have forsaken me the fountain of living waters, and hewed them out cisterns . . . that can hold no water." Or, as it is in Jeremiah xvii. 13, "they have forsaken the

Lord, the fountain of living waters." It was these living waters that our Lord referred to in St. John's Gospel when He was making it clear to the woman at the well that He was the fountain of living water.

It is no wonder that Naaman praised the river of Abana and its tributary Pharpar in 2 Kings v. 12, because this issues from the rocks in the Anti-Lebanon and is then absorbed in the Plain of Damascus, rendering it remarkably luxuriant. One can say that Damascus has properly one great river only, the Abanias as it is called today; but actually it divides up into seven streams which pass through various quarters of the city, carrying an abundant supply of water, not only for the gardens but for drinking purposes too.

The River Jordan, which Field Marshal Naaman looked down on, was a tiny little stream and a rather muddy one at that, compared to the great Abana. It was the River Jordan, however, in which John the Baptist carried out his baptisms; it is for this reason that the river is sometimes referred to as "the beginner of new life."

The main rivers of Scripture may be said to be the great river of Egypt, Genesis xv. 18, found on the west boundary of the Promised Land and now called the Nile; the River Euphrates which went along the east boundary of the Promised Land, and mentioned in Genesis ii. 14 and Joshua xxiv. 2; the River Jabbok which was the natural boundary between Moab and Ammon, Genesis xxxii. 22, now called the Zurka, and the River Kanah which bordered the inheritance of Ephraim and Manasseh as in Joshua xvi. 8.

In addition there were what may be called brooks or small streams, such as the Kedron, which our Lord and the Apostles crossed on the night of His betrayal (John

xviii. 1); the Kishon where the prophets of Baal were slaughtered and later Sisera was defeated in Judges iv. 7 and Judges v. 21; and the brook Zared which formed the boundary between Edom and Moab, now called, I believe, the El Ahsy (Numbers xxi. 12).

The other rivers mentioned are those found in Genesis ii which formed the boundary of paradise. One, Hiddekel, is said to be the Tigris, but the others, Pison and Gohon, are unidentified. It will thus be seen that Palestine was very fortunate in the amount of "living water" that was available.

It was impossible to grow grapes without ample water, and we see this in Ezekiel xvii 7: "He might water it by the furrows of her plantation. It was planted in a good soil by great waters, that it might bring forth branches, and that it might bear fruit". If, however, it was impossible to keep the furrows filled with water, then of course the vine would die. Ezekiel xvii. 10 says "it shall wither in the furrows where it grew." It is no good the market gardener setting out furrows, for there is no magic in this particular method of cultivation. There must be the living water available to fill the channels.

One has only to turn to Deuteronomy viii. 7 to note how God promised that the land of Palestine should be a country which could be irrigated quite easily. The actual words are "a good land, a land of brooks of water, of fountains and depths that spring out of valleys and hills . . ." It "is not as the land of Egypt . . . where thou sowedst thy seed, and wateredst it with thy foot".

Once again we see the emphasis on the value of the living water coming from the rivers and streams, as against the stored water used greatly by the Egyptians. Those, like the writer, who have travelled through the desert will know the joy of the aerated water of a bubbling brook, as compared with the somewhat stale flat taste of the water found in an oasis. The Jew was indeed fortunate in having so much living water available to him, as compared to the canal water which he previously had been used to in Egypt.

It must not be forgotten that the Children of Israel sojourned, as the Bible has it, in Egypt for four hundred and thirty years. That is a very long time. It is no wonder that during that period they multiplied from a few brothers to six hundred thousand men, plus of course the women and children. All they had known for all these years were the methods adopted by the Egyptians to irrigate the land. We learn from history that an equable system of irrigation was adopted by the Egyptians, as they tried to cope with the overflowing of the Nile year by year.

As Joseph, when he was Regent, managed to buy up all the land of Egypt during the famine for Pharaoh, it was possible to carry out what today would be called collective farming. This would not be appreciated by the Israelites who loved their independence. It must therefore have been a tremendous thing to leave Egypt where the water was rationed and where there was some regimented irrigation system, in order to go into a country where there was plenty of "living water" and where they could apply methods to ensure that their crops had all the water they needed.

CONCLUSION

THE GLORY OF GOD is declared in all the world of nature we see around us. It is His handiwork. The Psalmist says that the handiwork of God tells us something about His personality. Perhaps that is why the Son of God turned so readily to nature as He did so often when He talked about God.

An example of the Lord's easy use of the natural world to make a particular point about man and about God is the parable of the Vine in John 15: 'As the branch cannot bear fruit of itself, except it abide in the vine; no more can ye, except ye abide in Me. I am the vine, ye are the branches; he that abideth in Me and I in Him, the same bringeth forth much fruit; for without Me ye can do nothing."

Sometimes people talk about nature, forgetting that it is God's nature. It is His way of doing things. His personality is seen in nature. As the works of an artist bear clues that reveal the artist's feeling, the natural realm informs us, just a little bit, as to what God is like.

There is the element of return in nature . . . of life and growth and of decay and then the return of the elements to the earth . . . the evaporation of water and then the creation of clouds to form rain . . . all testifying to a trend in nature which can be called the complete circle. The passing back—the passing on—the ensuring that nothing is wasted. These facts have something to say about God . . . what He wants . . . what He wants for us.

In the creation is seen the declaration of the nature of the Creator. After all, why talk about weeds that grow in wheat-fields, or about mustard seeds, about vines and branches, and about fruit-bearing and branch-burning, if they do not have something to tell us? As the Lord used them, they tell us something about God and about man.

As our Lord told the simple story of the vine and the branches, He really emphasized the word "juncture" which is a fact of nature. You see it everywhere. Grafting and budding, for instance, causes perfect junctures. Look at the room you live in. Wall joins wall; corner joins the wall with the ceiling; the floor supports the corners at their point of connection.

As the Lord put it, branches have got to be attached to the stem of the vine, not only that they may produce but that they may live at all. Nothing lives in isolation. Nothing can exist all by itself. That's the way things are in nature, says the Old Book. The branch, if it is separated from its trunk, cannot survive. It ultimately dies by separation.

Thousands of years have gone by and still men make the same mistakes. It is true to say that it seems as if man cannot do otherwise, all by himself. The branch cannot bear fruit of itself, except it abide in the vine. When are we going to learn that? When are we going to learn that God wants every man to be in communion with Him? He wants, in fact, to establish an enduring, life-communicating, productive relationship with every man.

The branch cannot bear fruit itself, said Jesus Christ, except it abide in the vine. "No more can you," He added, "except you abide in Me. I am the vine, you are the branches. He that abideth in Me and I in him the same bringeth forth much fruit; for without Me you can do nothing." Christ is the joint to hold things together, to give the "structure form" which will hold up. Christ is not just another figure on the pages of history. He is God's Son, come into the world to put things right. He is the link between God and man *and* the mediator between God and man.

There has to be a connection from the leaf to the stem, from the stem to the trunk, from the trunk to the root, from the root to the pod, from the pod to the seed, and from the seed back to the plant again. Once again, the complete circle!

There has to be a connection from wheel to axle, from gear to gear, or from atom to atom, if the vehicle is to move. If the world is to have peace and order there has to be a connection between father and mother, parent and child, man and his world, and between the world and God. There has to be a connection between you and God. That connection is Jesus Christ. This once again is "The Complete Circle."

"If a man abide not in Me, he is cast forth as a branch and is withered; and men gather them, and cast them into the fire, and they are burned."